Automation, Collaboration, & E-Services

Volume 5

Series editor

Shimon Y. Nof, PRISM Center and School of Industrial Engineering, Purdue
University, West Lafayette, Indiana, USA
e-mail: nof@purdue.edu

About this Series

The Automation, Collaboration, & E-Services series (ACES) publishes new developments and advances in the fields of Automation, collaboration and e-services; rapidly and informally but with a high quality. It captures the scientific and engineering theories and techniques addressing challenges of the megatrends of automation, and collaboration. These trends, defining the scope of the ACES Series, are evident with wireless communication, Internetworking, multi-agent systems, sensor networks, and social robotics – all enabled by collaborative e-Services. Within the scope of the series are monographs, lecture notes, selected contributions from specialized conferences and workshops.

More information about this series at http://www.springer.com/series/8393

Rodrigo Reyes Levalle

Resilience by Teaming in Supply Chains and Networks

 Springer

Rodrigo Reyes Levalle
Operations Research and Advanced Analytics
American Airlines
Fort Worth, TX
USA

ISSN 2193-472X   ISSN 2193-4738   (electronic)
Automation, Collaboration, & E-Services
ISBN 978-3-319-86376-4   ISBN 978-3-319-58323-5   (eBook)
DOI 10.1007/978-3-319-58323-5

Printed on acid-free paper

This Springer imprint is published by Springer Nature
The registered company is Springer International Publishing AG
The registered company address is: Gewerbestrasse 11, 6330 Cham, Switzerland

To Mariel, Lara, and Agustina

Foreword

Resilience is on the mind and subconscious of all of us, as a means for survival. Defined many ways, it is commonly viewed as the transformability and adjustability, the ability to survive unforeseen circumstances, risks, or disruptions, and prevail despite high-impact events.

To survive, we are taught from an early age that if we fall or fail, we must get up and move on (walks, climbs, bicycles, horses, mistakes, tests, business, or social failures); that if a door is slammed in our face, we must look for a window of opportunity that will open. Resilience is also related to sustainability, and both are now considered important design objectives, particularly for the design of interconnected, inter-dependent, complex systems, and networked services.

The purpose of this book is to explain the meaning of resilience and its design in the broad context, and with a focus on the design and management of supply chains and supply networks. Written by Dr. Reyes Levalle, an experienced supply chains designer and supply networks engineer, the book is intended for beginners and advanced professionals, students, designers, policy makers, and managers. It is a pioneering effort to base resilience engineering and management on CCT, the collaborative control theory and tools.

What are those surprises, unexpected and high-impact events for which we have to be ready and resilient as individuals, societies, companies, and enterprises? Of course, we wish for only pleasant surprises, but consider the current news on meteors that are on their way to collide with our planet. They range from "city destroyers" to "planet destroyers," and we are reminded of how they disrupted the age of the dinosaurs on planet Earth. We are informed that it is taking a collaborative global effort to prepare for it, from monitoring and detection to design of preventive measures. Closer and more frequent disruptions are earthquakes, weather calamities, storms, floods, strikes, accidents, failures, energy loss, wars, and so on. For all, we need to prepare and design for resiliency, as explained systematically in this book.

Dr. Reyes Levalle explains clearly and authoritatively the concept of resilience and its significance, and how it applies to complex systems, supply networks, and supply chains. Applying a graph-based formalism, he then describes the fundamentals of disruption and response; restoration and adaptation; continued operation with acceptable quality of service; and scanning, detection, and preparedness. Design strategies along several dimensions of resilience follow, including structure, local and global, and flow control issues. Major innovative and original contributions by the author include flow control algorithms and protocols, resilience measurement, and the entire design of resilience by teaming.

Resilience involves active detection and prognosis of conflicts and errors potentially leading to harm and disruptions, and implementation of preventive/preparedness actions.

Based on the CCT principle of Fault Tolerance by Teaming (FTT), the *Resilience by Teaming* design framework is developed, including *sourcing protocols*, such as Sourcing Team Formation/Re-configuration Protocol (STF/RP) and Sourcing Flow Control Protocol (SFCP); *internal resource protocols*, such as Internal Flow Control Protocol (IFCP); and *delivery protocols*, such as Distribution Network Formation/Re-configuration Protocol (DNF/RP) and Distribution Flow Control Protocol (DFCP). Several detailed and validated case studies illustrate well how the framework and its associated protocols have been applied effectively to achieve the goal of design for resilience.

The author concludes with several open questions for future research in this area, concerning teaming-based resilience in supply network design and teaming-based protocols for resilient supply network operation. A general challenge is design rationalization—how agile and capable is the design in responding to alterations in information, processes, and systems' requirements? Those questions are a challenge, but this book already provides a clearly significant progress in resilience engineering and a major contribution to those who are responsible for supply resilience.

October 2016   Shimon Y. Nof
                                           Professor and Director, PRISM Center
                                                      Purdue University
                                                 West Lafayette, IN, USA

Contents

1	**Introduction**		1
	1.1	Complexity and Resilience in Supply Networks	1
	1.2	The Need for a Novel Approach: *Teaming*	3
	1.3	Book Summary	3
	References	4
2	**Supply Networks**		5
	2.1	Supply Chains... or Supply Networks?	5
	2.2	Supply Networks: An Introductory Definition	8
		2.2.1	Supply Network Agents	8
		2.2.2	Supply Network Links	8
		2.2.3	Supply Network Definition	10
	2.3	Graph Theory-Based Formalism for SNs	10
		2.3.1	Supply Network Dynamics	11
	2.4	Examples	12
	References	14
3	**Resilience Fundamentals for Supply Networks**		19
	3.1	Resilience in the Context of Supply Networks	19
	3.2	Fundamentals of Resilience in Supply Networks	22
		3.2.1	Inherence and Emergence	22
		3.2.2	Disruption and Response	22
		3.2.3	Restoration and Adaptation	23
		3.2.4	Endurance with Acceptable Quality of Service (QoS)	23
		3.2.5	Scanning, Detection, and Preparedness	23
	3.3	Resilience Dimensions and Levels	24
		3.3.1	Dimension (1): Supply Network Structure	24
		3.3.2	Dimension (2): Supply Network Control Protocols	25
		3.3.3	Resilience Levels: Local and Global	26

3.4   A Short Note on Resilience Versus Sustainability   26
3.5   Summary and Outlook   27
References.   28

4   Strategies to Design Resilient Supply Network Structures   31
4.1   Introduction   31
4.2   Redundancy   32
    4.2.1   Agent-Level (or Local) Redundancy   33
    4.2.2   Network-Level (or Global) Redundancy   35
4.3   Excess Resources: Storage and Capacity Surplus   37
    4.3.1   Capacity Surplus   37
    4.3.2   Storage   38
4.4   Communication Network Efficiency   39
4.5   Summary and Outlook   40
References.   40

5   Flow Control Protocols for Resilient Supply Networks   45
5.1   Sourcing Protocols: Controlling Input Flow   45
    5.1.1   Multi-sourcing.   46
    5.1.2   Back-up Sourcing   48
    5.1.3   Emergency Sourcing   49
    5.1.4   The Role of Communication in Sourcing
            Protocols   50
5.2   Internal Control Protocols: Managing Resources
      at Agent Level   50
5.3   Distribution Protocols: Controlling Flow Delivery
      to Successors   52
5.4   The Need for Situation Awareness   53
5.5   Summary and Outlook   54
References.   55

6   Resilience by Teaming Framework   59
6.1   The Need for Resilience Through Collaboration
      and Teaming.   59
6.2   An Overview of the Resilience by Teaming
      Framework   60
References.   63

**7   Resilience by Teaming: Sourcing Network Design
    and Flow Management Protocols**   65
7.1   Lessons Learnt from the FTT Principle of CCT
      Applied to Sensor Networks.   65
7.2   Sourcing Team Formation/Re-configuration Protocol
      (STF/RP)   67

7.2.1   Formalism of Sourcing Teams.   67
7.2.2   Design of the Primary Sourcing Team.   68
7.2.3   The Effect of Correlation on Primary Sourcing
        Team Selection   70
7.2.4   Design of the Secondary Sourcing Team.   72
7.2.5   Sourcing Team Formation/Re-configuration
        Protocol (STF/RP).   72
7.3   Sourcing Flow Control Protocol (SFCP).   74
7.3.1   Sourcing from the Primary Team:
        ST1 Protocol   76
7.3.2   Sourcing from the Secondary Team:
        ST2 Protocol   78
References.   81

8   Resilience by Teaming: Internal Resource Network Design,
    Flow Management, and Resource Control Protocols   83
    8.1   Design of Internal Resource Networks for Resilience
          Through Resource Teaming   83
          8.1.1   Resource Parallelism: Effect on Throughput
                  Average and Variability.   84
          8.1.2   Storage Design Guidelines for Management
                  Under Teaming Protocols   86
    8.2   Internal Flow Control Protocol (IFCP)   88
    References.   90

9   Resilience by Teaming: Distribution Network Design
    and Flow Management Protocols.   93
    9.1   Distribution Dynamics in Supply Networks   93
    9.2   Distribution Network Formation/Re-configuration
          Protocol (DNF/RP).   94
    9.3   Distribution Flow Control Protocol (DFCP)   98
    References.   100

10 Case Study A: Internal Flow Control Protocol Applied
    to Unreliable Production Lines.   101
    10.1   Introduction   101
           10.1.1   Production Network Formalism.   102
    10.2   Collaborative Production Control in a Tissue Paper
           Production Network   103
    10.3   Design of Experiments   105
    10.4   Case Study Results and Discussion   108
    References.   110

**11   Case Study B: Network Formation and Flow Control
      Protocols Applied to Physical Distribution Networks** 113
      11.1   Application of DNF/RP and DFCP to a Small Parcel
             Delivery Network 114
      11.2   Design of Experiments 117
      11.3   Case Study Results and Discussion 119
      References. 123

**12   Case Study C: Beyond Agent-Level Benefits—The Effect
      of Resilience by Teaming on Network-Level Resilience** 125
      12.1   Network Formation and Re-Configuration Dynamics 126
             12.1.1   Selection of Predecessors. 126
             12.1.2   Selection of Successors 128
             12.1.3   Modeling Supply Network Formation
                      and Re-Configuration Dynamics 129
      12.2   Design of Experiments 131
      12.3   Case Study Results and Discussion 134
             12.3.1   Pre-disruption Analysis 134
             12.3.2   Post-disruption Analysis: Effect on Total QoS
                      and Total Cost of Flow 135
             12.3.3   Network Topology: Results and Discussion 137
      References. 142

**13   Final Remarks and Outlook for Teaming-Based Resilience
      in Supply Networks** 143
      13.1   The Need for Resilience Fundamentals. 143
      13.2   Resilience by Teaming Framework. 144
      13.3   Teaming-Based Resilience in Supply Network Design:
             Open Questions 146
      13.4   Teaming-Based Protocols for Resilient Supply Network
             Operation: Open Questions. 148
      13.5   Future Research Directions. 149
      References. 150

Chapter 1
Introduction

Resilience research and practice require encompassing principles, which will enable the convergence of concepts and approaches across disciplines. Furthermore, sustainable resilience requires design and operation methodologies which acknowledge that supply network participants are fault-prone; they can, and will, fail at some point in time. Supply networks must be capable of maintaining operations despite these failures, with minimal protection from under-utilized resources.

Acknowledging the above facts, this book introduces a discussion on central resilience concepts and presents a novel approach to designing and operating resilient supply networks based on teaming. The conceptual discussion of the initial chapters distills key fundamentals from seemingly dissimilar disciplines in supply network research and provides the foundation for future resilience research. The latter chapters of this book present a comprehensive framework, Resilience by Teaming, developed based on the notion that a team of weaker agents can outperform a single flawless agent, under the right conditions. Combined, fundamentals and framework have the power to shape future complex supply networks to provide higher service levels with minimal disruption by enabling smart collaboration among fault-prone agents.

1.1 Complexity and Resilience in Supply Networks

Supply networks (SNs) are a collection of autonomous agents with self-interested goals that interact to enable physical, digital or service flow through a series of links. SN agents' interactions constitute a form of e-work, defined by Nof (2003) as *any collaborative, computer-supported and communication-enabled productive*

© Springer International Publishing AG 2018
R. Reyes Levalle, *Resilience by Teaming in Supply Chains and Networks*,
Automation, Collaboration, & E-Services 5, DOI 10.1007/978-3-319-58323-5_1

activities in highly distributed organizations of humans and/or robots or autonomous systems. Increasingly, e-work systems are augmenting their scale and becoming more complex. With a growing number of participants and interrelations, negative interactions among components are harder to prognose, anticipate, avoid, and/or recover from. Hence, these systems are progressively more exposed to unforeseen and/or unavoidable disruptions which affect performance of individual agents as well as larger collections thereof.

Resilience in the context of supply networks is an emerging concept related to the ability of agents, and the supply network itself, to handle disruptive events without losing performance. The concept is receiving increasing attention within complex adaptive systems research, in an effort to deepen our understanding of resilience-enabling design strategies and operation protocols. However, silo analysis of resilience in digital, physical, and service supply networks has prevented researchers to leverage from each other's findings in order to derive underlying resilience principles common to all supply networks.

Current frameworks typically focus on enabling SN agents' resilience but fail to address the emergence of SN resilience from decentralized agent-level decisions. At agent level, resilience arises from adaptive flow control protocols, available resources, and association/dissociation decisions that dynamically define agent interconnections. However, any local benefits obtained cannot be directly mapped to an improvement in SN resilience. Conversely, approaches producing marginal improvement on agent-level resilience may lead to significant increases in SN resilience; nonetheless, unless analyzed from a global perspective, these are likely to be discarded.

In recent years, research in complex systems has successfully linked network formation/re-configuration phenomena, driven by agent-level association/dissociation decisions, to emergent topology, network performance, and survivability (Barabási and Albert 1999; Albert et al. 2000; Albert and Barabási 2002; Tangmunarunkit et al. 2002; Thadakamalla et al. 2004; Brede and de Vries 2009). However, agent decision mechanisms analyzed rely on simple probabilistic association/dissociation rules and fail to account for more complex decision criteria normally involved in supply network formation and re-configuration processes.

Among current methodologies to enable agent-level resilience in SNs, most are based on trade-offs between quality of service (QoS) and the number/type of resources used. Designing and operating SNs with increased tied-up resources, in the form of storage, excess capacity, and/or higher predecessor reliability, enables greater fault-tolerance; however, as SNs become more complex, these approaches affect the network's long-term sustainability. Conversely, streamlining existing SNs' operations to attain higher efficiency in the use of resources may lead to networks where the effects of disruptions are more difficult to avoid or mitigate. Therefore, a new approach to creating agent-level and SN resilience, not based on the above mentioned trade-offs, is required.

1.2 The Need for a Novel Approach: *Teaming*

Nature and humans show us that it is possible to create resilient systems with fault-prone agent and resources through smart designs and distributed control protocols. The driving force behind these resilient systems is *Teaming*, a process by which a set of agents form a network and collaborate to achieve their individual goals and, perhaps, a common objective.

Over the last decades, several researchers have collaborated to develop and refine a set of six principles to design e-work systems, leading to the emergence of Collaborative Control Theory (Nof 2007). Although each of the Collaborative Control Theory (CCT) principles can have a meaningful impact on the design and control of resilient SNs, the principle of collaborative fault-tolerance or Fault-tolerance by Teaming (FTT) stands out as a potential enabler of resilient performance among agents susceptible to disruptions.

FTT principle is based on the notion that a team of weaker agents can outperform single flawless agent by enabling smart automation to overcome (temporarily) faulty agents (Nof 2007; Velásquez and Nof 2009). In this sense, protocols capable of teaming weaker agents to collectively overcome errors and conflicts, thus avoiding disruptions, appear as a promising solution to reduce the need for additional resources.

Inspired by the FTT principle of CCT, Chaps. 6–9 present a framework to achieve resilience in supply networks by creating teams of agents and enabling smart communication and collaboration. The novel approach yields significant performance improvements in operational parameters and costs, as shown by three real-world applications of Resilience by Teaming design guidelines and protocols, discussed in Chaps. 10–12.

1.3 Book Summary

The book is primarily organized in two main parts, in line with the discussions of the preceding sections. The first part, Chaps. 2–5, introduces the fundamentals to model supply networks and understand their resilience at different levels and from multiple dimensions. These theoretical foundations will guide future research efforts in physical, digital, and service supply networks resilience.

The second part, Chaps. 6–12 presents the Resilience by Teaming framework, a set of protocols and design guidelines by which supply network agents can dynamically form teams to achieve their individual goals in spite of their potential flaws, and illustrates the application thereof to real-world supply networks.

Finally, Chap. 13 summarizes the main contributions of this book and defined future research directions to extend the concepts and framework presented in this work.

References

Albert, R., & Barabási, A.-L. (2002). Statistical mechanics of complex networks. *Reviews of Modern Physics, 74*(1), 47–97. doi:10.1103/RevModPhys.74.47

Albert, R., Jeong, H., & Barabási, A.-L. (2000). Error and attack tolerance of complex networks. *Nature, 406*(6794), 378–382. doi:10.1038/35019019

Barabási, A.-L., & Albert, R. (1999). Emergence of scaling in random networks. *Science, 286* (5439), 509–512. doi:10.1126/science.286.5439.509

Brede, M., & de Vries, B. J. M. (2009). Networks that optimize a trade-off between efficiency and dynamical resilience. *Physics Letters A, 373*(43), 3910–3914. doi:10.1016/j.physleta.2009. 08.049

Nof, S. Y. (2003). Design of effective e-Work: Review of models, tools, and emerging challenges. *Production Planning & Control, 14*(8), 681–703. doi:10.1080/09537280310001647832

Nof, S. Y. (2007). Collaborative control theory for e-Work, e-Production, and e-Service. *Annual Reviews in Control, 31*(2), 281–292. doi:10.1016/j.arcontrol.2007.08.002

Tangmunarunkit, H., Govindan, R., Jamin, S., Shenker, S., & Willinger, W. (2002). Network topology generators: Degree-based vs. structural. *ACM SIGCOMM Computer Communication Review, 32*(4), 147–159.

Thadakamalla, H. P., Raghavan, U. N., Kumara, S., & Albert, A. (2004). Survivability of multiagent-based supply networks: A topological perspective. *IEEE Intelligent Systems, 19*(5), 24–31. doi:10.1109/MIS.2004.49

Velásquez, J. D., & Nof, S. Y. (2009). Collaborative e-Work, e-Business, and e-Service. In S. Y. Nof (Ed.), *Springer handbook of automation.* Springer: Berlin, pp. 1549–1576. doi:10.1007/978-3-540-78831-7_88

Chapter 2
Supply Networks

Real-world systems are inherently complex, formed by a large number of self-governing interconnected agents which dynamically update their behavior rules and connections based on context and environment changes. In order to model and understand the basic properties of these systems, it is necessary to start from a common definition of their components and main evolution processes.

In this chapter, the foundations introduced by Reyes Levalle and Nof (2017) to model supply networks as complex adaptive systems are presented. The traditional (linear) supply chain model is discussed and contrasted with a more advanced formalism, i.e., supply networks. Supply network components, agents and links, as well as basic network structure dynamics, formation and reconfiguration, are defined based on the notion of autonomous agents and graph theory. The concepts introduced in the following sections will provide the theoretical basis for supply network modelling and analysis, and for the discussion of supply network resilience, throughout this book and in future research efforts in the field.

2.1 Supply Chains… or Supply Networks?

By definition, a chain is a collection of elements that are connected to each other forming a line; each element is connected to, at most, two other elements. Initially, this notion of sequential connections provided a sound conceptual model for real-world manufacturing processes entailing the transformation of natural resources into finished products, giving rise to the idea of a Supply Chain (SC). Over time, researchers and practitioners introduced various definitions of SC, e.g.:

© Springer International Publishing AG 2018
R. Reyes Levalle, *Resilience by Teaming in Supply Chains and Networks*,
Automation, Collaboration, & E-Services 5, DOI 10.1007/978-3-319-58323-5_2

(1) An integrated process to source, manufacture, and deliver products involving forward product and backward information flows (Beamon 1998)
(2) A sequence of production and distribution activities (Stevenson and Spring 2007)
(3) A system of firms that are linked via buyer-supplier relationships (Dass and Fox 2011).

Despite the increase in SCs' complexity, with growing number of participants and interrelations, the concept of SC as a sequence of stages is still in place, as evidenced, for instance, by the current version of the Supply Chain Operations Reference model (Luck and D'Inverno 2001; Huan et al. 2004; Supply Chain Council 2010; Leukel and Sugumaran 2013). This widely used framework, developed by the Supply Chain Council to model SCs, proposes a sequential model where a company will focus on its suppliers, internal processes, and customers, and may eventually include more distant participants upstream or downstream (e.g., the company suppliers' suppliers).

Formally, a Supply Chain can be defined as:

> A set of agents arranged in sequential stages or, more traditionally, echelons, where physical flow occurs in one direction between connected agents in neighboring echelons, and information flow occurs in the opposite direction.

The model is product (flow)-centric (Braziotis et al. 2013) and/or agent-centric (Christopher and Peck 2004) in that the SC will contain enough information to fully describe the flow of one type of product or the flow relative to the transformation made by one agent. This representation of the interrelations among agents presents three main shortcomings: (a) incompleteness, (b) intransitivity, and (c) non-reversible flows.

Agents can be part of several SCs (Braziotis et al. 2013), nonetheless, connectivity of agents, beyond the focus of the SC under consideration, is not adequately represented (Dass and Fox 2011). This incomplete representation of the interactions that are relevant to the SC focus in order to model the competition over upstream resources may lead to sub-optimal, or even adverse, decisions. For instance, consider Fig. 2.1a where the SC of agent A is depicted by grey circles in echelons −2 to +1.

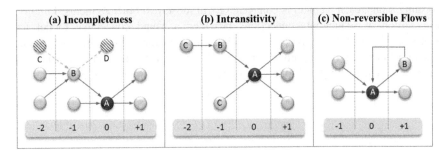

Fig. 2.1 Main shortcomings of the supply chain concept

Agent B is a direct predecessor of A and is also part of a different SC with agents C and D. Although D and A may not compete over the same market, they are in direct competition for the resources of B and, therefore, the existence of a connection between B and D is relevant to A. Moreover, C is a predecessor of B whose actions can affect the decisions B makes regarding how it allocates resources to serve A and D (e.g., if C reduces the cost of a raw material B used to provide D, it could make it more desirable for B to allocate more capacity to D and less to A) and, therefore, it is of relevance to A.

Intransitivity refers to the inability of the SC model to include transitive relations among agents, where C is a predecessor of B and A, and B is a predecessor of A, as shown in Fig. 2.1b. By constraining the SC agents to belong to sequential echelons and flow to occur between neighboring echelons, transitive relations must be modeled by replicating the predecessor common to two or more agents, potentially overlooking the effects of correlation. For instance, in Fig. 2.1b a disruption in C will directly affect A, but it will also affect B which, in turn will affect A.

The notion of sequential flow between neighboring stages also constraints the modeling of reverse flows, necessary for re-manufacturing and recycling as well as for returns of flow that did not met the service level agreement (SLA) between two agents. These needs led to the relaxation of the sequential flow constraint and gave rise to the concept of "closed-loop" SCs as an extension of traditional SCs (see Stindt and Sahamie (2012) for a review on closed-loop SCs).

Although the SC model could be further relaxed and modified to include new extensions and overcome some of its inherent limitations, the introduction of a new paradigm to describe relationships among agents, free from the SCs shortcomings, seems more suitable. As argued by Fiksel (2003), it is not possible to model complex non-linear systems by connecting a series of fragmented linear models; these models oversimplify and distort the reality of real-world systems (Hearnshaw and Wilson 2013).

Lately, researchers have been increasingly embracing the notion of Supply Network (SN) to describe the complex interrelationships among agents (Lamming et al. 2000; Harland et al. 2001; Christopher and Peck 2004; Thadakamalla et al. 2004; Pathak et al. 2007; Stevenson and Spring 2007; Dass and Fox 2011; Nair and Vidal 2011; Corominas 2013; Hearnshaw and Wilson 2013); nevertheless, there is no unified definition for SNs. Flow in SNs is not necessarily restricted to physical goods; it can also involve services (Lamming et al. 2000; Harland et al. 2001; Cui et al. 2010; Corominas 2013; Hearnshaw and Wilson 2013) or tasks (Moghaddam et al. 2016). Moreover, the supply of digital flow, e.g., signals, digital information, and/or data in communication and sensor networks, can also be described under the paradigm of a SN.

Despite the differences among physical, digital, and service SNs (or any combination thereof) it is possible to develop a unifying framework for their design, operation, and performance analysis. The advantage resides in that the framework will leverage the complementarity of previous research focused solely on physical, service, or digital SNs; care must be taken to evaluate whether certain assumptions pertaining one network type invalidate the use of some approaches from a different domain or, conversely, to examine which changes need to be made to method-ologies proven effective in one domain to also be beneficial in other domains.

2.2   Supply Networks: An Introductory Definition

In general, a supply network can be defined as a collection of interconnected agents that interact via communications and flow exchanges to fulfill their individual objectives and/or a set of common goals. In order to formalize a definition and model for SNs that encompasses physical, digital, and service SNs, it is required first to characterize its two principal components: SN agents and SN links.

2.2.1   Supply Network Agents

Lately, engineering and science research has been increasingly focusing on prob-lems concerning the interaction of multiple entities which make decisions with some degree of decentralization. The advent of multi-agent systems theory has led to the introduction of various definitions of the term *agent*; nonetheless, several attributes and components are common to most variants. Agents, which can be physical (e.g., humans, robots, sensors, manufacturing facilities) or intangible (e.g., software programs) entities, must be (1) autonomous—able to control their own actions and internal resources, (2) goal-oriented—capable of making rational decisions leading towards the achievement of their own goals, and (3) interactive—able to communicate with other agents and to react to changes in the context or environment (Franklin and Graesser 1997; Luck and D'Inverno 2001; Nof 2003; Allwood and Lee 2005; Velásquez and Nof 2009; Wooldridge and Jennings 2009).

Based on the aforementioned characteristics of agents, a SN agent (a), Fig. 2.2, can be defined as:

> An autonomous entity with self-interested goals comprising (1) a set of internal resources R_a capable of processing input into output and (2) a collection of control protocols CP_a which coordinate collaboration among internal resources and regulate communication and interactions with other agents.

2.2.2   Supply Network Links

SN agents interact to enable physical (e.g., products), digital (e.g., signals), and/or service (e.g., tasks) flow among their internal resources and to exchange informa-tion via control protocols. Flows and information exchanges form unidirectional

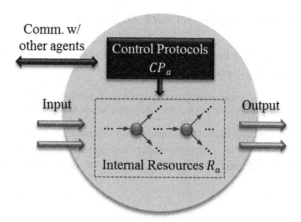

Fig. 2.2 Supply network agent

associations, i.e., in the direction of flow or communication, between SN agents; these associations can be formally characterized as links.

A flow link ($fl_{i \rightarrow j}$) is formally defined as:

A connection between the internal resources of SN agents i and j which enables flow from i to j.

Every flow link has a set of time-dependent attributes $\Omega[fl_{i \rightarrow j}]$, e.g., capacity, status, which fully describe the interaction between i and j at time t.

Communication links, on the other hand, are part of the SN control structure and enable interaction between two agents. Formally, a communication link ($cl_{i \rightarrow j}$) is:

A connection between the control protocols of SN agents i and j which enables information to be sent from i to j.

As in the case of flow links, every communication link has a set of time-dependent attributes $\Omega[cl_{i \rightarrow j}]$, e.g., capacity, status, which fully describe the interaction between i and j at time t.

Whenever two agents i and j form an association, the direction of the link $i \rightarrow j$ determines precedence; therefore, i is a predecessor of i and, conversely, i is a successor of i. Each link between any two agents i and i comprises a service level agreement ($SLA_{i \rightarrow j}$), in which, limits and valid ranges for relevant flow and communication link attributes are defined. Then, at any time t, the quality of service provided to agent i by agent i, $QoS_{i \rightarrow j}$ can be expressed as a function of $SLA_{i \rightarrow j}$, $\Omega[fl_{i \rightarrow j}]$, and $\Omega[cl_{i \rightarrow j}]$.

$$QoS_{i \rightarrow j}(t) = f\left(SLA_{i \rightarrow j}, \Omega[fl_{i \rightarrow j}], \Omega[cl_{i \rightarrow j}]\right).$$

2.2.3 Supply Network Definition

Agents and links are the building blocks of Supply Networks (SNs), which can be formally defined as:

> A collection of autonomous agents with self-interested goals that interact to enable physical, digital or service flow through a series of links.

SN agents' interactions constitute a form of e-work, defined by Nof (2003) as any collaborative, computer-supported and communication-enabled productive activities in highly distributed organizations of humans and/or robots or autonomous systems.

2.3 Graph Theory-Based Formalism for SNs

Consider a set $A = \{1, \ldots, a\}$ of supply network agents which form associations by establishing flow and communication links, and let $FL = \{fl_{i \to j} | i, j \in A \text{ and } i \neq j\}$ and $CL = \{cl_{i \to j} | i, j \in A \text{ and } i \neq j\}$ be the set of flow links and communication links, respectively. Then, from a mathematical perspective, a supply network SN can be represented by a directed graph with nodes $a \in A$ and arcs $l \in L = FL \cup CL$. Lee II et al. (2007) use a similar approach to model interdependent network flows by introducing a layer (network) for each type of flow and adding constraints on flows at nodes common to several networks to account for interdependences. It can be argued, though, that each layer in the interdependent network is either a flow or communication network and that interdependency constraints are part of the SN agents' control protocols CP_a.

Based on the links formed by agents in A, the latter can be classified as source (or input) agents, kernel agents, and sink (or output) agents. A source agent $a \in A^I \subseteq A$ is a SN agent that lacks input flow links. Formally,

$$A^I = \{a \in A | \not\exists fl_{i \to a} \in FL, i \in A\}$$

A sink agent $a \in A^O \subseteq A$ is a SN agent that lacks output flow links. Formally,

$$A^O = \{a \in A | \not\exists fl_{a \to j} \in FL, j \in A\}$$

A kernel agent $a \in A^K \subseteq A$ is a SN agent which has at least one input and one output flow link with other agents in A. Formally,

$$A^K = \{a \in A | \exists fl_{i \to a} \in FL \text{ and } \exists fl_{a \to j} \in FL, \quad i, j \in A\}$$

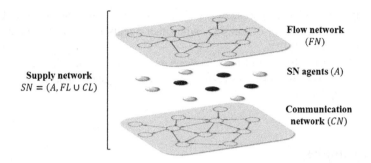

Fig. 2.3 Supply networks: a combination of flow and communication networks

Therefore, from a functional viewpoint, a supply network $SN = (A, L)$ can be formally defined as:

The conjunction of a flow network $FN = (A, FL)$ and a communication network $CN = (A, CL)$ through which agents in A interact in order to fulfill their individual objectives and/or a set of common goals (Fig. 2.3). Interactions enable input from source agents $A^I \subset A$ to flow through kernel agents $A^K \subset A$ and reach sink agents $A^O \subset A$.

The objective of the SN is to provide sink agents in A^O with an adequate QoS based on their individual service level agreement (SLA) with agents either in A^K or A^I. Then, quality of service at network level can be measured by aggregating local QoS values as follows

$$QoS_{SN}(t) = f\left(SLA_{i \to j}, \Omega[fl_{i \to j}], \Omega[cl_{i \to j}]\right), \quad \forall i \in A^I \cup A^K \text{ and } j \in A^O.$$

2.3.1 Supply Network Dynamics

Several authors have studied the mechanisms behind network evolution over time from a structural perspective (the reader is referred to Albert and Barabási (2002) for a summary of previous research in this area). Despite the variety of processes that can shape a SN's structure, these can be clustered in two main categories: formation and re-configuration.

Formation is the process by which a supply network $SN^* = (A^*, L^*)$ is obtained from $SN = (A, L)$ by (1) adding a set of agents A^+ and links L^+ to, and/or (2) removing a set of agents A^- and links L^- from SN.

Formally, let A, A^+ and A^- be sets of agents such that $A^+ \cap A = \emptyset$ and $A^- \subseteq A$; then, $A^* = A^+ \cup (A \cap A^{-c})$. Moreover, let L^+ be the set of links between agents i and j such that $i \in A$ and $j \in A^+$, or $j \in A$ and $i \in A^+$, and let L^- be the set of links between h and k such that $L^- \subseteq L$ where at least one element in $\{h, k\} \in A^-$. Then, $L^* = L^+ \cup (L \cap L^{-c})$

Reconfiguration is the process by which a set of agents in A create a set of links L^+ and/or eliminate a set of links L^- with other agents in A, transforming $SN = (A, L)$ into $SN^* = (A, L^*)$.

Formally, let L^+ be a set of links between agents $i, j \in A$ such that $L^+ \cap L = \phi$ and L^- be a set of links between agents $h, k \in A$ such that $L^- \subseteq L$; then, $L^* = L^+ \cup (L \cap L^{-c})$.

2.4 Examples

The formalism introduced in the previous section is independent of the real-world realization of agents, their objectives, the type of information exchanged, and the flow/s which are central to the supply network; therefore, it can be used to conceptualize and model physical, digital, and service supply networks. Table 2.1 presents a summary of the examples described in the next paragraphs along with the main findings of this section.

Digital supply networks involve a set of interconnected agents responsible for generating, storing, transforming, transmitting, and receiving signals or information, e.g., sensor networks (Colbourn 1993; Liu and Nof 2004; Jeong and Nof 2009; Erdene-Ochir et al. 2010), and computer and communication networks (Colbourn 1993; Faloutsos et al. 1999; Tangmunarunkit et al. 2002; Sabella and Iovanna 2006; Lau et al. 2008; Pereira and Neumann 2009). In sensor networks, agents include sensing devices, computational processors of intelligent sensors, base stations, and external processing units, and, links, both for flow and communication, comprise physical or wireless connections among agents. Similarly, in computer and communication networks, e.g., Internet or a LAN, agents involve routing devices, servers, databases, user interfaces, and field devices (sensors and actuators), and links include all physical, e.g., cable, fiber optic, and wireless, e.g., radio frequency, Wi-Fi, communication channels. The common characteristic among digital supply networks is that FN and CN share a single physical/wireless network, i.e., communication messages among agents are transmitted through the same physical network used for signal transmission.

In physical supply networks, agents are also responsible for generating, storing, transforming, delivering, and receiving flow, but in the form of physical products. Physical SN agents include (but are not limited to) production equipment, manufacturing facilities, warehouses, power generation plants, pumping stations, water treatment plants, cross-docking facilities, and ports, among other. Examples of physical SNs include production and manufacturing networks (Paquet et al. 2004, 2008; Adenso-Diaz et al. 2012; Hu et al. 2013), logistics and transportation networks (Meepetchdee and Shah 2007; Kaluza et al. 2010; Baroud et al. 2014), and utility networks, e.g., water (Todini 2000) and power generation/distribution grids (Rosas-Casals 2010; Zio and Piccinelli 2010; Zio and Sansavini 2013).

Table 2.1 Summary of supply network examples

SN domain	Agents (A) and links (L)	Example(s)
Physical	**A:** Equipment/facilities that generate, store, transform, and deliver, flow of physical products **L:** Conveyor belts, roads, pipes	Production networks (Paquet et al. 2004, 2008; Adenso-Diaz et al. 2012; Hu et al. 2013) Transportation networks (Meepetchdee and Shah 2007; Kaluza et al. 2010; Baroud et al. 2014) Utility networks (Todini 2000; Rosas-Casals 2010; Zio and Piccinelli 2010; Zio and Sansavini 2013
Digital	**A:** Devices that generate, store, transform, transmit, and receive signals or information **L:** Cables, optic fiber, radio frequencies	Sensor networks (Colbourn 1993; Liu and Nof 2004; Jeong and Nof 2009; Erdene-Ochir et al. 2010) Computer and communication networks (Colbourn 1993; Faloutsos et al. 1999; Tangmunarunkit et al. 2002; Sabella and Iovanna 2006; Lau et al. 2008; Pereira and Neumann 2009)
Service	**A:** Resources that execute tasks, e.g., operators, software agents, processing units **L:** Logic sequence of steps required to execute a workflow	Product development projects (Lee and Suh 2008) Workflow management systems (Cao et al. 2004) Collaborative networked organizations (Moghaddam et al. 2016)
Combined	**A, L:** Any combination of physical, digital, and service agents/links	Equipment maintenance (Mahulkar et al. 2009) Power generation/distribution (Lee II et al. 2007) Integrated business process management and manufacturing networks (Luo et al. 2001; Dotoli et al. 2006; Zhang et al. 2010)

Service networks involve the flow of tasks between agents which are responsible for their execution. Upon receiving a task from a predecessor, an agent must process it using its internal resources. Once processing is completed, an agent can send additional tasks to other agents within the service network. Workflow networks, e.g., product development projects (Lee and Suh 2008) and software workflow management (Cao et al. 2004), in which activities or tasks need to be executed by various agents to complete a project or request are examples of service networks.

Oftentimes, physical supply networks need to interface with service networks (Cui et al. 2010) or other physical networks. In these interactions, the primary physical network requires specific flow from the secondary physical network or the execution of tasks or allocation of resources from a service network. Examples of these interactions are (1) equipment maintenance, where a crew is responsible for performing maintenance tasks on the resources of a set of processes which are part of a physical supply network (Mahulkar et al. 2009), (2) power generation/ distribution, in which a set of agents is responsible for providing the physical supply network agents with energy to function (Lee II et al. 2007), and integrated business process management and manufacturing networks, where business management agents make decisions affecting the operation of manufacturing agents (Luo et al. 2001; Dotoli et al. 2006; Zhang et al. 2010).

References

Adenso-Diaz, B., Mena, C., Garcia-Carbajal, S., & Liechty, M. (2012). The impact of supply network characteristics on reliability. *Supply Chain Management, 17*(3), 263–276. doi:10.1108/13598541211227108

Albert, R., & Barabási, A.-L. (2002). Statistical mechanics of complex networks. *Reviews of Modern Physics, 74*(1), 47–97. doi:10.1103/RevModPhys.74.47

Allwood, J. M., & Lee, J.-H. (2005). The design of an agent for modelling supply chain network dynamics. *International Journal of Production Research, 43*(22), 4875–4898. doi:10.1080/00207540500168295

Baroud, H., Ramirez-Marquez, J. E., Barker, K., & Rocco, C. M. (2014). Stochastic measures of network resilience: Applications to waterway commodity flows. *Risk Analysis: An Official Publication of the Society for Risk Analysis, 34*(7), 1317–1335. doi:10.1111/risa.12175

Beamon, B. M. (1998). Supply chain design and analysis: Models and methods. *International Journal of Production Economics, 55*(3), 281–294. doi:10.1016/S0925-5273(98)00079-6

Braziotis, C., Bourlakis, M., Rogers, H., & Tannock, J. (2013). Supply chains and supply networks: Distinctions and overlaps. *Supply Chain Management: An International Journal, 18*(6), 644–652. doi:10.1108/SCM-07-2012-0260

Cao, J., Wang, J., Zhang, S., & Li, M. (2004). A dynamically reconfigurable system based on workflow and service agents. *Engineering Applications of Artificial Intelligence, 17*(7), 771–782. doi:10.1016/j.engappai.2004.08.030

Christopher, M., & Peck, H. (2004). Building the resilient supply chain. *International Journal of Logistics Management, 15*(2), 1–14. doi:10.1108/09574090410700275

Colbourn, C. J. (1993). Analysis and synthesis problems for network resilience. *Mathematical and Computer Modelling, 17*(11), 43–48. doi:10.1016/0895-7177(93)90251-S

Corominas, A. (2013). Supply chains: What they are and the new problems they raise. *International Journal of Production Research, 51*(23–24), 6828–6835. doi:10.1080/00207543.2013.852700

Cui, L., Kumara, S., & Albert, R. (2010). Complex networks: An engineering view. *IEEE Circuits and Systems Magazine, 10*(3), 10–25. doi:10.1109/MCAS.2010.937883

Dass, M., & Fox, G. L. (2011). A holistic network model for supply chain analysis. *International Journal of Production Economics, 131*(2), 587–594. doi:10.1016/j.ijpe.2011.01.025

Dotoli, M., Fanti, M. P., & Meloni, C. (2006). Design and optimization of integrated E-supply chain for agile and environmentally conscious manufacturing. *IEEE Transactions on Systems, Man, and Cybernetics—Part A: Systems and Humans, 36*(1), 62–75. doi:10.1109/TSMCA.2005.859189

Erdene-Ochir, O., Minier, M., Valois, F., and Kountouris, A., 2010. Resiliency of wireless sensor networks: Definitions and analyses, in: *2010 IEEE 17th International Conference on Telecommunications.* pp. 828–835. Doha, Qatar, doi:10.1109/ICTEL.2010.5478822

Faloutsos, M., Faloutsos, P., & Faloutsos, C. (1999). On power-law relationships of the Internet topology. *ACM SIGCOMM Computer Communication Review, 29*(4), 251–262. doi:10.1145/316194.316229

Fiksel, J. (2003). Designing resilient, sustainable systems. *Environmental Science and Technology, 37*(23), 5330–5339. doi:10.1021/es0344819

Franklin, S., and Graesser, A., 1997. Is it an agent, or just a program?: A taxonomy for autonomous agents, in: *Proceedings of ECAI'96 Workshop (Intelligent Agents III Agent Theories, Architectures, and Languages)*. Springer, Budapest, Hungary, pp. 103–115. doi:10.1007/BFb0013568

Harland, C. M., Lamming, R. C., Zheng, J., & Johnsen, T. E. (2001). A taxonomy of supply networks. *The Journal of Supply Chain Management, 37*(4), 21–27. doi:10.1111/j.1745-493X.2001.tb00109.x

Hearnshaw, E. J. S., & Wilson, M. M. J. (2013). A complex network approach to supply chain network theory. *International Journal of Operations & Production Management, 33*(4), 442–469. doi:10.1108/01443571311307343

Hu, Y., Li, J., & Holloway, L. E. (2013). Resilient control for serial manufacturing networks with advance notice of disruptions. *IEEE Transactions on Systems, Man, and Cybernetics: Systems, 43*(1), 98–114. doi:10.1109/TSMCA.2012.2189879

Huan, S. H., Sheoran, S. K., & Wang, G. (2004). A review and analysis of supply chain operations reference (SCOR) model. *Supply Chain Management: An International Journal, 9*(1), 23–29. doi:10.1108/13598540410517557

Jeong, W., & Nof, S. Y. (2009). A collaborative sensor network middleware for automated production systems. *Computers & Industrial Engineering, 57*(1), 106–113. doi:10.1016/j.cie.2008.11.007

Kaluza, P., Kölzsch, A., Gastner, M. T., & Blasius, B. (2010). The complex network of global cargo ship movements. *Journal of the Royal Society, Interface/the Royal Society, 7*(48), 1093–1103. doi:10.1098/rsif.2009.0495

Lamming, R., Johnsen, T., Zheng, J., & Harland, C. (2000). An initial classification of supply networks. *International Journal of Operations & Production Management, 20*(6), 675–691. doi:10.1108/01443570010321667

Lau, C. H., Soong, B.-H., & Bose, S. K. (2008). Preemption with rerouting to minimize service disruption in connection-oriented networks. *IEEE Transactions on Systems, Man, and Cybernetics—Part A: Systems and Humans, 38*(5), 1093–1104. doi:10.1109/TSMCA.2008.2001075

Lee, E. E., II, Mitchell, J. E., & Wallace, W. A. (2007). Restoration of services in interdependent infrastructure systems: A network flows approach. *IEEE Transactions on Systems, Man and Cybernetics, Part C (Applications and Reviews), 37*(6), 1303–1317. doi:10.1109/TSMCC.2007.905859

Lee, H., & Suh, H.-W. (2008). Estimating the duration of stochastic workflow for product development process. *International Journal of Production Economics, 111*(1), 105–117. doi:10.1016/j.ijpe.2007.01.003

Leukel, J., & Sugumaran, V. (2013). Formal correctness of supply chain design. *Decision Support Systems, 56*, 288–299. doi:10.1016/j.dss.2013.06.008

Liu, Y., & Nof, S. Y. (2004). Distributed microflow sensor arrays and networks: Design of architectures and communication protocols. *International Journal of Production Research, 42*(15), 3101–3115. doi:10.1080/00207540410001699363

Luck, M., & D'Inverno, M. (2001). A conceptual framework for agent definition and development. *The Computer Journal, 44*(1), 1–20. doi:10.1093/comjnl/44.1.1

Luo, Y., Zhou, M., & Caudill, R. J. (2001). An integrated e-supply chain model for agile and environmentally conscious manufacturing. *IEEE/ASME Transactions on, Mechatronics, 6*(4), 377–386. doi:10.1109/3516.974851

Mahulkar, V., McKay, S., Adams, D. E., & Chaturvedi, A. R. (2009). System-of-systems modeling and simulation of a ship environment with wireless and intelligent maintenance technologies. *IEEE Transactions on Systems, Man, and Cybernetics Part A: Systems and Humans, 39*(6), 1255–1270. doi:10.1109/TSMCA.2009.2025140

Meepetchdee, Y., & Shah, N. (2007). Logistical network design with robustness and complexity considerations. *International Journal of Physical Distribution & Logistics Management, 37*(3), 201–222. doi:10.1108/09600030710742425

Moghaddam, M., Nof, S. Y., & Menipaz, E. (2016). Design and administration of collaborative networked headquarters. *International Journal of Production Research, 54*(23), 7074–7090. doi:10.1080/00207543.2015.1125544

Nair, A., & Vidal, J. M. (2011). Supply network topology and robustness against disruptions—An investigation using multi-agent model. *International Journal of Production Research, 49*(5), 1391–1404. doi:10.1080/00207543.2010.518744

Nof, S. Y. (2003). Design of effective e-Work: Review of models, tools, and emerging challenges. *Production Planning & Control, 14*(8), 681–703. doi:10.1080/09537280310001647832

Paquet, M., Martel, A., & Desaulniers, G. (2004). Including technology selection decisions in manufacturing network design models. *International Journal of Computer Integrated Manufacturing, 17*(2), 117–125. doi:10.1080/09511920310001593100

Paquet, M., Martel, A., & Montreuil, B. (2008). A manufacturing network design model based on processor and worker capabilities. *International Journal of Production Research, 46*(7), 2009–2030. doi:10.1080/00207540600821009

Pathak, S. D., Dilts, D. M., & Biswas, G. (2007). On the evolutionary dynamics of supply network topologies. *IEEE Transactions on Engineering Management, 54*(4), 662–672. doi:10.1109/TEM.2007.906856

Pereira, C.E., and Neumann, P., 2009. Industrial communication protocols, in: Nof, S.Y. (Ed.), *Springer Handbook of Automation*. pp. 981–999. Springer, Berlin, doi:10.1007/978-3-540-78831-7_56

Reyes Levalle, R., & Nof, S.Y. (2017). Resilience in supply networks: Definition, dimensions, and levels. *Annual Reviews in Control, 43*, pp. 224–236, ISSN 1367-5788,doi: 10.1016/j.arcontrol.2017.02.003

Rosas-Casals, M., 2010. Power grids as complex networks: Topology and fragility, in: *Proceedings of the 2010 Complexity in Engineering*. pp. 21–26. Rome, Italy, doi:10.1109/COMPENG.2010.23

Sabella, R., & Iovanna, P. (2006). Self-adaptation in next-generation internet networks: How to react to traffic changes while respecting QoS? *IEEE Transactions on Systems, Man and Cybernetics, Part B (Cybernetics), 36*(6), 1218–1229. doi:10.1109/TSMCB.2006.873210

Stevenson, M., & Spring, M. (2007). Flexibility from a supply chain perspective: Definition and review. *International Journal of Operations & Production Management, 27*(7), 685–713. doi:10.1108/01443570710756956.

Stindt, D., & Sahamie, R. (2012). Review of research on closed loop supply chain management in the process industry. *Flexible Services and Manufacturing Journal, 26*(1–2), 268–293. doi: 10.1007/s10696-012-9137-4

Supply Chain Council, 2010. Supply chain operations reference (SCOR) model—overview.

Tangmunarunkit, H., Govindan, R., Jamin, S., Shenker, S., & Willinger, W. (2002). Network topology generators: Degree-based vs. structural. *ACM SIGCOMM Computer Communication Review, 32*(4), 147–159.

Thadakamalla, H. P., Raghavan, U. N., Kumara, S., & Albert, A. (2004). Survivability of multiagent-based supply networks: A topological perspective. *IEEE Intelligent Systems, 19*(5), 24–31. doi:10.1109/MIS.2004.49

Todini, E. (2000). Looped water distribution networks design using a resilience index based heuristic approach. *Urban Water, 2*(2), 115–122. doi:10.1016/S1462-0758(00)00049-2

Velásquez, J.D., and Nof, S.Y., 2009. Collaborative e-Work, e-Business, and e-Service, in: Nof, S. Y. (Ed.), *Springer Handbook of Automation*. pp. 1549–1576. Springer, Berlin, doi:10.1007/978-3-540-78831-7_88

Wooldridge, M., & Jennings, N. R. (2009). Intelligent agents: Theory and practice. *The Knowledge Engineering Review, 10*(2), 115. doi:10.1017/S0269888900008122

Zhang, Y., Huang, G. Q., Qu, T., & Ho, O. (2010). Agent-based workflow management for RFID-enabled real-time reconfigurable manufacturing. *International Journal of Computer Integrated Manufacturing, 23*(2), 101–112. doi:10.1080/09511920903440354

Zio, E., & Piccinelli, R. (2010). Randomized flow model and centrality measure for electrical power transmission network analysis. *Reliability Engineering & System Safety, 95*(4), 379–385. doi:10.1016/j.ress.2009.11.008

Zio, E., & Sansavini, G. (2013). Vulnerability of smart grids with variable generation and consumption: A system of systems perspective. *IEEE Transactions on Systems, Man, and Cybernetics: Systems, 43*(3), 477–487. doi:10.1109/TSMCA.2012.2207106

Chapter 3
Resilience Fundamentals for Supply Networks

Several authors recognize the work of (Holling 1973) in ecosystems' resilience as the seminal work in systems resilience. Based on the observed behavior and evolution of various ecological systems, Holling (1973) defined resilience as a property thereof, responsible for the persistence of relationships within the ecosystem in the face of changes to system variables or parameters. Over the next 40 years, research in systems resilience spread over a wide range of disciplines, from psychology to supply network management; nevertheless, resilience in the context of supply networks still remains largely unaddressed (Bhamra et al. 2011).

In this chapter, the definition and components of resilience in the context of supply networks introduced in Reyes Levalle and Nof (2017) are analyzed. Five fundamental principles of supply network resilience are introduced, in line with resilience definitions from several supply network domains. These core concepts will provide a platform to unify resilience understanding across domains in supply network research. Furthermore, in order for supply network principles to materialize, resilience must be addressed in two dimensions, structure and control protocols, and at two levels, agent-level and network-level. These resilience components are introduced in the subsequent sections and discussed within the supply network formalism presented in Chap. 2.

3.1 Resilience in the Context of Supply Networks

Most of the research efforts related to supply network resilience comprise defining the term "resilience" in the context of a supply network, understanding its components and dimensions, and formalizing the processes leading to resilience. Despite the fact that the definition of resilience is still being molded, various authors

© Springer International Publishing AG 2018
R. Reyes Levalle, *Resilience by Teaming in Supply Chains and Networks*,
Automation, Collaboration, & E-Services 5, DOI 10.1007/978-3-319-58323-5_3

Table 3.1 Summary of resilience definitions and their domain of application

Article	Resilience definition	SN domain[a]		
		P	S	D
Todini (2000)	Capability of the designed system to react and to overcome stress conditions	X		
Ta et al. (2009)	Ability of a transportation system to absorb the consequences of disruptions, reduce the impact thereof, and maintain freight mobility	X		
Hu et al. (2013)	Ability of a system or enterprise to minimize the effects of a disruption	X		
Rice and Caniato (2003)	Ability of an organization to react to an unexpected disruption and restore normal operations	X	X	
Christopher and Peck (2004)	Ability of a system to return to its original state or move to a new, more desirable state after being disturbed	X	X	
Sheffi (2005)	Being better positioned than competitors to deal with, and even gain advantage from, disruptions	X	X	
Fiksel (2006)	Capacity for an enterprise to survive, adapt, and grow in the face of turbulent change	X	X	
Tang (2006)	Property of a firm's strategy; to sustain operations during a major disruption and quickly recover after it occurs	X	X	
Falasca et al. (2008)	Ability of a supply chain to reduce (1) the probabilities of disruptions, (2) the consequences thereof, and (3) the time to recover normal performance	X	X	
Ponomarov and Holcomb (2009)	Adaptive capability to prepare for unexpected events, respond to disruptions, and recover from them by maintaining continuity of operations at desired levels of connectedness and control over structure and function	X	X	
Reed et al. (2009)	Ability to *bounce back* after a major disturbance	X	X	
Pettit et al. (2010)	Capacity for an enterprise to survive, adapt, and grow in the face of turbulent change [after Fiksel (2006)]	X	X	
Antunes (2011)	Capability to maintain operations under a wide spectrum of potential breakdowns	X	X	
Burnard and Bhamra (2011)	Emergent property of organizational systems related to the inherent and adaptive qualities and capabilities that enable adaptive capacity during turbulent periods	X	X	
Mamouni Limnios et al. (2014)	The magnitude of disturbance a system can tolerate and still persist [after Gunderson and Holling (2002)]	X	X	
Spiegler et al. (2012)	Adaptive capability to prepare for unexpected events, respond to disruptions, and recover from them by maintaining continuity of operations at desired levels of connectedness and control over structure and function [after Ponomarov and Holcomb (2009)]	X	X	
Hearnshaw and Wilson (2013)	Extent to which a system is able to perform its functions despite disruptions or damage created by disturbances	X	X	
Gong et al. (2014)	Ability to recover quickly from disruptions and ensure customers are minimally affected	X	X	

(continued)

Table 3.1 (continued)

Article	Resilience definition	SN domain[a]		
		P	S	D
Najjar and Gaudiot (1990)	Maximum number of node failures that can be sustained while the network remains connected with a probability $1 - p$			X
Colbourn (1993)	Expected number of node pairs which can communicate in the network, when links fail independently with known probabilities			X
Brede and de Vries (2009)	Ability to recover a previous operational state after a dynamic perturbation			X
Erdene-Ochir et al. (2010)	Ability of a network to continue to operate in presence of k compromised nodes; the capacity of a network to endure and overcome internal attacks			X
Sterbenz et al. (2010, 2011a, 2011b)	Ability of a network to provide and maintain an acceptable level of service in the face of various faults and challenges to normal operation			X
Smith et al. (2011)	Ability of a network to defend against, and maintain an acceptable level of service in the presence of, malicious attacks, software and hardware faults, human mistakes and large-scale natural disasters			X
Fiksel (2003)	Capacity of a system to tolerate disturbances while retaining its structure and function	X	X	X
Sheffi and Rice (2005)	Ability to *bounce back* from a disruption	X	X	X
Henry and Ramirez-Marquez (2012)	Ability of an entity to recover from an external disruptive event	X	X	X
Pant et al. (2014)	Ability of systems to *bounce back* from disruptive events, be they malevolent attacks, man-made accidents, or natural disasters	X	X	X

[a]*P* Physical, *S* Service, *D* Digital

propose different frameworks to create and/or achieve resilience in supply networks; their approaches diverge in several directions and vaguely overlap, in part, because of the lack of a unified resilience definition.

In order to characterize the notion of resilience for physical, service, and digital supply networks, definitions from research articles covering a broad range of disciplines, from supply chain management to communication networks literature, need to be analyzed. Table 3.1 presents a summary of the definitions found in a survey of relevant articles.

3.2 Fundamentals of Resilience in Supply Networks

Based on an extensive survey and analysis of resilience definitions, Reyes Levalle (2015) introduces five resilience fundamentals. These core concepts are engrained in most definitions found in a wide range of supply network domains, and will guide future supply network resilience definitions and applications.

3.2.1 Inherence and Emergence

> Resilience is an inherent ability of SN agents and/or an emergent ability of supply networks.

In most cases, authors refer to resilience as an ability of an agent or network, or use a synonym or closely related word, e.g., capacity, capability, or property, suggesting resilience is a built-in capacity of agents and/or SNs. The dual level at which resilience can manifest is not explicitly stated by the reviewed definitions, but it is addressed by Reed et al. (2009) and Burnard and Bhamra (2011) and can be inferred from cross comparison, e.g., system versus entity from "ability of a system to return to its original state or move to a new, more desirable state after being disturbed" (Christopher and Peck 2004) and "ability of an entity to recover from an external disruptive event" (Henry and Ramirez-Marquez 2012). The dual level is also evidenced in reviews encompassing different disciplines such as physics, ecology, psychology, and disaster management (Bhamra et al. 2011).

3.2.2 Disruption and Response

> Resilience is related to the responses which follow the occurrence of a disruption to normal operation by undesired (but not necessarily unforeseen) events.

Most of the definitions reviewed suggest that resilience drives the response that follows undesired events which affect flow and/or communications, e.g., stress, disturbances, disruptions, turbulent change/periods, failures, attacks. In other words, if a supply network is free from disruptions, resilience cannot be observed.

Undesired events can be (i) random, as implied by the use of words such as failure or disturbance, and/or reference to probability of occurrence (Najjar and Gaudiot 1990; Colbourn 1993; Falasca et al. 2008; Sterbenz et al. 2010, 2011a, b; Hearnshaw and Wilson 2013; Hu et al. 2013), or (ii) targeted, as it can be inferred from the presence of words such as attack (Erdene-Ochir et al. 2010; Smith et al. 2011; Pant et al. 2014). Furthermore, disruptive events can be unexpected (Ponomarov and Holcomb 2009; Spiegler et al. 2012) or predictable, as suggested by the possibility of being better positioned for (Sheffi 2005), or defending against (Smith et al. 2011), such events.

3.2.3 Restoration and Adaptation

Resilience involves restoring Quality of Service to a, possibly new, stable, normal state.

Recovery from disruptions always involves a restoration process in which a set of mechanisms return an agent or a SN to a normal pre-disruption state, thus restoring output conditions. This process is suggested in various definitions by the use of verbs such as overcome, return, recover, survive, bounce back, or persist (Todini 2000; Rice and Caniato 2003; Sheffi and Rice 2005; Fiksel 2006; Reed et al. 2009; Gong et al. 2014; Mamouni Limnios et al. 2014; Pant et al. 2014).

Nevertheless, restoration may not suffice in some cases. When a previously unknown disruption occurs, SN structure and control mechanisms may need to be adapted in order to be able to anticipate and resolve future occurrences of the, now known, disruption. Christopher and Peck (2004), Fiksel (2006), and Pettit et al. (2010) suggest that the recovery process may lead to a different state by adjusting some characteristic/s of the topology and/or protocols so as to minimize or eliminate the effect of a given disruption, should it occur again.

3.2.4 Endurance with Acceptable Quality of Service (QoS)

Resilience involves maintaining acceptable Quality of Service (QoS) from the occurrence of a disruption until restoration to a stable, normal state is achieved.

The goal of maintaining acceptable QoS is directly stated in resilience definitions from digital SNs (Erdene-Ochir et al. 2010; Sterbenz et al. 2010, 2011a, b; Smith et al. 2011) whereas in physical/service SNs authors focus on maintaining continuity of operations (Fiksel 2003; Tang 2006; Ponomarov and Holcomb 2009; Ta et al. 2009; Antunes 2011; Spiegler et al. 2012). It can be argued, nevertheless, that the purpose of maintaining operations is directly related to the objective of retaining as much QoS as possible following a disruptive event. Although not explicit, it is implied that the recovery process from (3.2.3) should not bring the QoS below the acceptable QoS level during the transition to a stable, normal state.

3.2.5 Scanning, Detection, and Preparedness

Resilience involves active detection and prognosis of conflicts and errors potentially leading to disruptions, and implementation of preventive/preparedness actions.

Reviewed definitions implicitly assume that agents and SNs are capable of detecting disruptions whenever they occur. Moreover, early detection and prognosis of conflicts and errors potentially leading to disruptions is central to resilience, as it enables agents and SNs to anticipate and prepare for such events (Nof 2007; Chen and Nof 2009). This concept is evidenced in phrases such as "be better positioned" (Sheffi 2005), "reduce probabilities and consequences of disruptions" (Falasca et al. 2008), "prepare for unexpected events" (Ponomarov and Holcomb 2009; Spiegler et al. 2012), and "defend against disruptions" (Smith et al. 2011).

3.3 Resilience Dimensions and Levels

Although it is not explicitly addressed in the reviewed definitions of resilience, several authors, e.g., Rice and Caniato (2003), Christopher and Peck (2004), Sheffi and Rice (2005), and Jackson and Ferris (2013) discuss the need for design for resilience. This requirement is also embedded in the notions of mitigation and contingency from Tomlin (2006) and Spiegler et al. (2012), which call for a combination of structural characteristics and control strategies to reduce the impact of disruptions and enable quick recovery. Sterbenz et al. (2011a) suggest that, in order to design for resilience, each SN agent must first define its service level requirements (SLAs) and come to a SLA with its predecessors. It follows from the design principles and concepts in the reviewed articles that there are two main dimensions involved in SN resilience: structure, i.e., topology and resources, and control protocols, and two levels of resilience: local and global, i.e., at agent level and network level, respectively. These dimensions and levels are not to be addressed in isolation; their interaction is fundamental in enabling resilience and, therefore, it is critical to optimize the match between SN structure and the control protocols used to manage flow and communication [e.g., as shown in the analysis of transportation networks in Ta et al. (2009)]

3.3.1 Dimension (1): Supply Network Structure

The structure of a SN comprises its topology and the resources available at each agent. SN topology relates to the characteristics of the network $SN = (A, L)$ emerging from the interconnections formed by links L among SN agents $a \in A$; the concept extends to also include topography, i.e., the physical location of SN agents. SN resources refers to the collection of agent resources R_a, $a \in A$ available in a given $SN = (A, L)$. The contribution of resources e.g., production and storage capacity, to resilience is straightforward; the larger the amount of available resources compared to the level required for normal operation, the more resilient a SN will be (Sheffi 2005). Similarly, it can be argued that higher interconnection may lead to increased resilience through augmented visibility and the existence of multiple alternatives. Nevertheless, both resources and topology involve one-time costs e.g., purchase, connection, or installation, and operational costs e.g., maintenance and energy, and therefore need to be designed within certain constraints not to affect SN sustainability.

Several authors address the relevance of topological features to resilience of networks. In the context of communication networks, Sterbenz et al. (2011a) state that in order to evaluate the resilience of a network it is necessary to assess its physical topology and geography first, as these determine the ability to withstand disruptions. Albert et al. (2000) and Albert and Barabási (2002) provide insights on how node/link removal affects networks with topologies emerging from different formation mechanisms, and address the effects of allowing the network to re-configure, i.e., to restore eliminated links and/or create new ones. Najjar and Gaudiot (1990) analyze the performance of computer networks with regular graph topologies under random failures and conclude that large-scale networks with a constant node degree, i.e., number of links per agent, are less resilient to random failures than smaller networks. On the other hand, Brede and de Vries (2009) show that kernel agents degree is reduced and size of A^K increases as the network is designed for increased resilience while maintaining short paths between input agents A^I and sink agents A^O.

Erdene-Ochir et al. (2010) study the performance of routing protocols in wireless sensor networks (WSNs) and conclude that connectivity, i.e., the existence of several alternative connection paths between input agents A^I and sink agents A^O, and high node degree are requirements for resilience. In addition to connectivity and node degree, Sterbenz et al. (2010) point out the need for redundancy, i.e., replication of agents and/or links, to provide increased fault-tolerance. Nonetheless, redundancy must be designed taking into account geographical location of SN agents and links, to avoid failure correlation (Sterbenz et al. 2011a).

The relation of SN topology and resilience is also discussed among authors in physical and service SNs literature. Hearnshaw and Wilson (2013) highlight the importance of topological features of the network to achieve "efficient supply chains". Craighead et al. (2007) and Falasca et al. (2008) argue that node density, a measure of geographical proximity among nodes, is important for resilience, introduces the need for topography considerations. Thadakamalla et al. (2004) suggest that for some applications it is also necessary to account for the interaction between network topology and nodes' capabilities, as adding links without careful analysis thereof may not lead to higher resilience.

3.3.2 Dimension (2): Supply Network Control Protocols

Flow management in SNs requires control protocols CP_a that address three aspects: agent-level control of internal operations, distributed control of multiple agents via coordination and negotiation protocols, and emergent SN behavior. As pointed out by Erdene-Ochir et al. (2010) in their analysis of the ability of different routing protocols for WSNs to withstand and overcome attacks, there is a dearth of literature addressing the resilience of control protocols. The article also analyzes the importance of certain topological characteristics of the SN (e.g., connectivity and node degree) to the capacity of protocols to be resilient (e.g. through the presence of multiple alternative routing paths). Similarly, Sterbenz et al. (2011a) discuss the need for both resilient network structure and resilient protocol layers in communication networks. Cholda et al. (2007) point out that not all sink agents in A^O require the same level of resilience, and survey protocols for digital SNs that enable resilience differentiation.

In physical and service SNs, Gong et al. (2014) analyze the optimization of restoration protocols subject to the topological constraints of the network, in an attempt to define the best matching recovery protocols for each of the networks involved. The interplay between SN topology and control protocols for system resilience is also addressed by Pant et al. (2014) in an study of inland waterway port disruptions.

3.3.3 Resilience Levels: Local and Global

Resilience is an inherent ability of SN agents and/or an emergent ability of supply networks. This core concept clearly defines two levels of resilience: local, i.e., the inherent ability of SN agents to be resilient, and global, i.e., the emerging capability of a SN of being resilient. The dual level at which resilience can manifest is found both on in physical and digital SNs: Spiegler et al. (2012) conduct a review of models to evaluate resilience in physical SNs at a local level and conclude that they cannot be used to assess global resilience; Smith et al. (2011) discuss the need to incorporate defensive measures during the design of digital SNs at individual nodes and across the network domain.

The interplay between design for resilience, of both structure and control protocols, and the two levels of SN resilience has not been addressed in literature thus far. Further work is required to understand how local structure design strategies and control protocols impact global network resilience, and to analyze how network-level mechanisms of information sharing can aid local design to increase both local and global resilience.

3.4 A Short Note on Resilience Versus Sustainability

Although it is not addressed in the reviewed definitions of resilience, it is worth noting that several authors introduce and discuss a trade-off between cost and resilience. Sheffi (2005) observes that resilience can be achieved by creating redundancies and incorporating protection across a physical/service SN, e.g., multiple suppliers, high inventory, levels, low capacity utilization; however, this approach will inevitably lead to significant cost increases. Similarly, Todini (2000) argues that resilience in water networks can be increased by looping a fraction of the total water flow, at the expense of higher energy consumption. Christopher and Peck (2004) point out the need to re-evaluate the trade-off between efficiency and redundancy, and suggest that, despite the inevitable increase in short term costs, having excess resources will, to some extent, increase SN resilience. Spiegler et al. (2012) elaborates on the cost perspective by drawing attention to the opportunity costs that may arise from lacking resilience, e.g., lost revenue from poor customer service. Research in digital SNs also evidences the cost versus resilience trade-off. Cholda et al. (2007) propose that it is possible to select recovery schemes that enable faster restoration of QoS after a disruption, provided that the digital SN can incur in higher costs. Sterbenz et al. (2010, 2011a) generalizes the idea of cost to include also the use resources as part of the trade-off.

Interestingly, the combination of (i) use of resources, (i) actual versus opportunity costs, and (iii) short versus long term effects emerging from the reviewed articles suggest an interdependence between resilience and sustainability. However, only a few authors have commented on this interdependence: Fiksel (2003, 2006) describes resilience as the essence of sustainability, Nof (2013) argues that the challenge of resilience is to enable sustainability despite disruptions, and, in the field of ecological-social systems, Folke et al. (2002) review two case studies that show a connection between resilience and sustainability. Moreover, the literature survey produced only one article, i.e., Derissen et al. (2011) in the area of ecological-economic systems, in which the interdependence between resilience and sustainability is analyzed mathematically and, contrary to the aforementioned observations, the authors conclude that resilience and sustainability are independent concepts. Based on the scarcity of research addressing the interdependence between resilience and sustainability from a mathematical perspective and the contradicting results found thus far, the question of whether a SN can be resilient yet not sustainable or vice versa remains unanswered.

3.5   Summary and Outlook

The concept of resilience is pervasive, extending beyond SNs to organizational systems, psychology, ecology, and disaster management (Bruneau et al. 2003; Tierney and Bruneau 2007; Bhamra et al. 2011). Despite the vast range of research areas addressing resilience, and numerous works in the context of supply networks, the concept still lacks a unified definition. Although it is unlikely for the precise wording of the definition to be shared among authors, all definitions must encompass the five resilience fundamentals for SNs (Reyes Levalle 2015):

(1) Inherence and emergence
(2) Disruption and response
(3) Restoration and adaptation
(4) Endurance with acceptable QoS
(5) Scanning, detection, and preparedness.

Further work is required to map the abovementioned resilience fundamentals to resilience dimensions, levels, and design and control strategies, and to understand the relation between agent-level and network-level resilience. In this line, the SN formalism introduced in Chap. 2 as well as the discussions in subsequent chapters constitute a first attempt at formalizing and elucidating such interrelations. Deeper understanding of these aspects of SN resilience is key to formulate guidelines for design for resilience, which must address structural and control protocol dimensions at agent and network level.

Finally, as briefly discussed in the end of this chapter, there is a potential interrelation between resilience and sustainability that needs to be further explored by researchers in the coming years. Is resilience a necessary condition for sustainability? Is it a sufficient condition? At a first glance, resilience appears to be more related to shorter term SN dynamics while sustainability is more concerned with the long-term survivability of the SN. However, intuition suggests that there cannot be long-term SN survival without the capacity to overcome short-term disruptions. Work on this emerging area has still to prove the nature of the interrelation, if any.

References

Albert, R., & Barabási, A.-L. (2002). Statistical mechanics of complex networks. *Reviews of Modern Physics, 74*(1), 47–97. doi:10.1103/RevModPhys.74.47

Albert, R., Jeong, H., & Barabási, A.-L. (2000). Error and attack tolerance of complex networks. *Nature, 406*(6794), 378–382. doi:10.1038/35019019

Antunes, P. (2011). BPM and exception handling: Focus on organizational resilience. *IEEE Transactions on Systems, Man, and Cybernetics, Part C (Applications and Reviews), 41*(3), 383–392. doi:10.1109/TSMCC.2010.2062504

Bhamra, R., Dani, S., & Burnard, K. (2011). Resilience: The concept, a literature review and future directions. *International Journal of Production Research, 49*(18), 5375–5393. doi:10.1080/00207543.2011.563826

Brede, M., & de Vries, B. J. M. (2009). Networks that optimize a trade-off between efficiency and dynamical resilience. *Physics Letters A, 373*(43), 3910–3914. doi:10.1016/j.physleta.2009.08.049

Bruneau, M., Chang, S. E., Eguchi, R. T., Lee, G. C., O'Rourke, T. D., Reinhorn, A. M., et al. (2003). A framework to quantitatively assess and enhance the seismic resilience of communities. *Earthquake Spectra, 19*(4), 733–752. doi:10.1193/1.1623497

Burnard, K., & Bhamra, R. (2011). Organisational resilience: Development of a conceptual framework for organisational responses. *International Journal of Production Research, 49*(18), 5581–5599. doi:10.1080/00207543.2011.563827

Chen, X.W., & Nof, S.Y. (2009). Automating errors and conflicts prognostics and prevention. In S.Y. Nof (Ed.), *Springer handbook of automation* (pp. 503–525). Springer: Berlin. doi:10.1007/978-3-540-78831-7_30

Cholda, P., Mykkeltveit, A., Helvik, B., Wittner, O., & Jajszczyk, A. (2007). A survey of resilience differentiation frameworks in communication networks. *IEEE Communications Surveys & Tutorials, 9*(4), 32–55. doi:10.1109/COMST.2007.4444749

Christopher, M., & Peck, H. (2004). Building the resilient supply chain. *International Journal of Logistics Management, 15*(2), 1–14. doi:10.1108/09574090410700275

Colbourn, C. J. (1993). Analysis and synthesis problems for network resilience. *Mathematical and Computer Modelling, 17*(11), 43–48. doi:10.1016/0895-7177(93)90251-S

Craighead, C. W., Blackhurst, J., Rungtusanatham, M. J., & Handfield, R. B. (2007). The severity of supply chain disruptions: Design characteristics and mitigation capabilities. *Decision Sciences, 38*(1), 131–156. doi:10.1111/j.1540-5915.2007.00151.x

Derissen, S., Quaas, M. F., & Baumgärtner, S. (2011). The relationship between resilience and sustainability of ecological-economic systems. *Ecological Economics, 70*(6), 1121–1128. doi:10.1016/j.ecolecon.2011.01.003

Erdene-Ochir, O., Minier, M., Valois, F., & Kountouris, A. (2010). Resiliency of wireless sensor networks: Definitions and analyses. In *2010 IEEE 17th international conference on telecommunications* (pp. 828–835). Doha, Qatar. doi:10.1109/ICTEL.2010.5478822

Falasca, M., Zobel, C. W., & Cook, D. (2008). A decision support framework to assess supply chain resilience. In *Proceedings of the 5th international ISCRAM conference* (pp. 596–605). Washington, DC

Fiksel, J. (2003). Designing resilient, sustainable systems. *Environmental Science and Technology, 37*(23), 5330–5339. doi:10.1021/es0344819.

Fiksel, J. (2006). Sustainability and resilience: Toward a systems approach. *Sustainability: Science, Practice, & Policy, 2*(2), 14–21.

Folke, C., Carpenter, S., Elmqvist, T., Gunderson, L., Holling, C. S., & Walker, B. (2002). Resilience and sustainable development: Building adaptive capacity in a world of transformations. *Ambio, 31*(5), 437–440. doi:10.1639/0044-7447(2002)031[0437:RASDBA]2.0.CO;2

Gong, J., Mitchell, J. E., Krishnamurthy, A., & Wallace, W. A. (2014). An interdependent layered network model for a resilient supply chain. *Omega, 46*, 104–116. doi:10.1016/j.omega.2013.08.002

Gunderson, L. H., & Holling, C. S. (2002). *Panarchy: Understanding transformations in human and natural systems*. Washington, DC: Island Press.

Hearnshaw, E. J. S., & Wilson, M. M. J. (2013). A complex network approach to supply chain network theory. *International Journal of Operations & Production Management, 33*(4), 442–469. doi:10.1108/01443571311307343

Henry, D., & Ramirez-Marquez, J. E. (2012). Generic metrics and quantitative approaches for system resilience as a function of time. *Reliability Engineering & System Safety, 99*, 114–122. doi:10.1016/j.ress.2011.09.002

Holling, C. S. (1973). Resilience and stability of ecological systems. *Annual Review of Ecology and Systematics, 4*(1), 1–23. doi:10.1146/annurev.es.04.110173.000245

Hu, Y., Li, J., & Holloway, L. E. (2013). Resilient control for serial manufacturing networks with advance notice of disruptions. *IEEE Transactions on Systems, Man, and Cybernetics: Systems, 43*(1), 98–114. doi:10.1109/TSMCA.2012.2189879

Jackson, S., & Ferris, T. L. J. (2013). Resilience principles for engineered systems. *Systems Engineering, 16*(2), 152–164. doi:10.1002/sys.21228

Mamouni Limnios, E. A., Mazzarol, T., Ghadouani, A., & Schilizzi, S. G. M. (2014). The resilience architecture framework: Four organizational archetypes. *European Management Journal, 32*(1), 104–116. doi:10.1016/j.emj.2012.11.007

Najjar, W., & Gaudiot, J.-L. (1990). Network resilience: A measure of network fault tolerance. *IEEE Transactions on Computers, 39*(2), 174–181. doi:10.1109/12.45203

Nof, S. Y. (2007). Collaborative control theory for e-Work, e-Production, and e-Service. *Annual Reviews in Control, 31*(2), 281–292. doi:10.1016/j.arcontrol.2007.08.002

Nof, S.Y. (2013). Sustainability and resiliency in supply networks. In *Plenary Talk—14th Asia Pacific Industrial Engineering and Management Society*. Cebu City, Philippines.

Pant, R., Barker, K., Ramirez-Marquez, J. E., & Rocco, C. M. (2014). Stochastic measures of resilience and their application to container terminals. *Computers & Industrial Engineering, 70*, 183–194. doi:10.1016/j.cie.2014.01.017

Pettit, T. J., Fiksel, J., & Croxton, K. L. (2010). Ensuring supply chain resilience: Development of a conceptual framework. *Journal of Business Logistics, 31*(1), 1–21. doi:10.1002/j.2158-1592.2010.tb00125.x

Ponomarov, S. Y., & Holcomb, M. C. (2009). Understanding the concept of supply chain resilience. *International Journal of Logistics Management, 20*(1), 124–143. doi:10.1108/09574090910954873

Reed, D. A., Kapur, K. C., & Christie, R. D. (2009). Methodology for assessing the resilience of networked infrastructure. *IEEE Systems Journal, 3*(2), 174–180. doi:10.1109/JSYST.2009.2017396

Reyes Levalle, R. (2015). *Resilience by teaming in supply networks*. West Lafayette: Purdue University.

Reyes Levalle, R., & Nof, S. Y. (2017). Resilience in supply networks: Definition, dimensions, and levels. *Annual Reviews in Control, 43*, pp. 224–236, ISSN 1367-5788. doi: 10.1016/j.arcontrol.2017.02.003

Rice, J. B., Jr., & Caniato, F. (2003). Building a secure and resilient supply network. *Supply Chain Management Review, 7*(5), 22–30.

Sheffi, Y. (2005). Building a resilient supply chain. *Harvard Business Review, 1*(8), 2–6.

Sheffi, Y., & Rice, J. B., Jr. (2005). A supply chain view of the resilient enterprise. *MIT Sloan Management Review, 47*(1), 41–48.

Smith, P., Fessi, A., Lac, C., Hutchison, D., Sterbenz, J. P. G., Scholler, M., et al. (2011). Network resilience: A systematic approach. *IEEE Communications Magazine, 49*(7), 88–97. doi:10.1109/MCOM.2011.5936160

Spiegler, V. L. M., Naim, M. M., & Wikner, J. (2012). A control engineering approach to the assessment of supply chain resilience. *International Journal of Production Research, 50*(21), 6162–6187. doi:10.1080/00207543.2012.710764

Sterbenz, J. P. G., Cetinkaya, E. K., Hameed, M. A., Jabbar, A., Qian, S., & Rohrer, J. P. (2011a). Evaluation of network resilience, survivability, and disruption tolerance: analysis, topology generation, simulation, and experimentation. *Telecommunication Systems, 52*(2), 705–736. doi:10.1007/s11235-011-9573-6

Sterbenz, J. P. G., Cetinkaya, E. K., Hameed, M. A., Jabbar, A., & Rohrer, J. P. (2011b). Modelling and analysis of network resilience. In *Proceedings of the third international conference on communication systems and networks* (pp. 1–10). Bangalore. doi:10.1109/COMSNETS.2011.5716502

Sterbenz, J. P. G., Hutchison, D., Çetinkaya, E. K., Jabbar, A., Rohrer, J. P., Schöller, M., et al. (2010). Resilience and survivability in communication networks: Strategies, principles, and survey of disciplines. *Computer Networks, 54*(8), 1245–1265. doi:10.1016/j.comnet.2010.03.005

Ta, C., Goodchild, A. V., & Pitera, K. (2009). Structuring a definition of resilience for the freight transportation system. *Transportation Research Record: Journal of the Transportation Research Board, 2097*, 19–25. doi:10.3141/2097-03

Tang, C. S. (2006). Perspectives in supply chain risk management. *International Journal of Production Economics, 103*(2), 451–488. doi:10.1016/j.ijpe.2005.12.006

Thadakamalla, H. P., Raghavan, U. N., Kumara, S., & Albert, A. (2004). Survivability of multiagent-based supply networks: A topological perspective. *IEEE Intelligent Systems, 19*(5), 24–31. doi:10.1109/MIS.2004.49

Tierney, K., & Bruneau, M. (2007). Conceptualizing and measuring resilience: A key to disaster loss reduction. *TR News, 250*, 14–17.

Todini, E. (2000). Looped water distribution networks design using a resilience index based heuristic approach. *Urban Water, 2*(2), 115–122. doi:10.1016/S1462-0758(00)00049-2

Tomlin, B. (2006). On the value of mitigation and contingency strategies for managing supply chain disruption risks. *Management Science, 52*(5), 639–657. doi:10.1287/mnsc.1060.0515

Chapter 4
Strategies to Design Resilient Supply Network Structures

The supply network formalism and resilience fundamentals introduced in Chaps. 2 and 3 provide the theoretical foundation to understand resilience in SNs. Practitioners, however, require actionable insights and strategies to create and operate resilient supply networks. Literature from physical, digital, and service supply networks contains numerous techniques to achieve these objectives. Nevertheless, the field lacks a clear mapping of these techniques to core resilience strategies to design and operate SNs.

In this chapter, the three main strategies applied in the design of resilient structures in SNs—redundancy, excess resources, communication network efficiency—are presented. The aim is to provide practitioners and researchers with a high-level map of design strategies which encompass many existing approaches, and, possibly, several future ones, to guide their efforts to create resilient SNs. Strategies are discussed in connection with existing examples found in literature to provide a wide overview of techniques used within each strategy to achieve resilience at agent and network level.

4.1 Introduction

As expressed by Christopher and Peck (2004), resilience should be designed in a SN; by purposely adding certain structural characteristics in combination with adequate control protocols it is possible to increase a SN's resilience. Sterbenz et al. (2011) argue that resilient SNs must be designed to incorporate passive defenses, i.e., structural conditions that reduce the probability/impact of disruptions on QoS, and active defenses, i.e., control protocols, that combined with the right structural properties provide recovery and adaptation capabilities. The authors discuss several resilience strategies and highlight that the right combination of strategies must be selected and located in the network so as to simultaneously optimize resilience and

© Springer International Publishing AG 2018
R. Reyes Levalle, *Resilience by Teaming in Supply Chains and Networks*,
Automation, Collaboration, & E-Services 5, DOI 10.1007/978-3-319-58323-5_4

cost (of operation and resilience mechanisms). Moreover, the work of Gong et al. (2014) on the interaction of structural conditions and recovery protocols stresses the importance of designing structural capacity and capabilities that can be leveraged by adequately chosen control protocols.

From the articles reviewed in Chap. 3, three main design strategies for resilient supply networks can be extracted: (1) redundancy, (2) excess resources, i.e., surplus capacity and storage, and (3) situation awareness and efficient communications. Redundancy and protection through excess resources aim at adding layers of structural protection against disruptions, thus enabling resilience. Nonetheless, these design strategies may entail significant installation and/or operation costs in physical, service, and digital SNs and need to be adequately allocated. Situation awareness capabilities and efficient information flow are required in order to supply preventive/preparedness control protocols with timely information to anticipate, and possibly avoid, disruptions by optimally managing and allocating available structural protection. The following sections discuss each of the design strategies in detail.

4.2 Redundancy

At a fundamental level, redundancy is related to the existence of alternatives capable of providing the same functionality but not required simultaneously to achieve a given objective. In general, authors discuss redundancy in a broad sense. As evidenced in Table 4.1, authors' definitions encompass topological redundancy, e.g., multiple connections to predecessors with equal function, and excess resources, e.g., inventory held for protection. Nonetheless, from a SN design perspective, it is more accurate to consider topological redundancy and excess resources as separate resilience-enabling strategies, based on the fact that each strategy is driven by a different process. Topological redundancy (henceforth referred to as redundancy) is built into the SN through formation and re-configuration processes, formalized in Sect. 2.3.1, in which the SN agents create/dissolve links with other SN agents. Conversely, excess resources are created by capacity sizing decisions, e.g., buffer allocation, and/or by the parameters chosen for control protocols, e.g., stocks and replenishment levels.

Redundancy is a pre-requisite, oftentimes tacit, for well-studied control protocols that leverage the availability of multiple connections, e.g., multi-path routing in digital SNs for increased message delivery under failures (Al-Karaki and Kamal 2004) and multiple supplier strategies in physical SNs to cope with leadtime variability and delivery disruptions (Minner 2003). Nevertheless, the mere existence of redundancy does not guarantee resilience; redundancy needs to be properly managed by adequately chosen control protocols in order to capitalize on its potential benefits (Ta et al. 2009; Erdene-Ochir et al. 2010).

Following the above observation, it is worth noting that protocols can exploit redundancy at agent level, i.e., multiple input/output links, and/or at network level, i.e., the existence of multiple paths between any two agents. In line with the two

Table 4.1 Summary of redundancy definitions and their domain of application

Article	Redundancy Definition	SN Domain[a]		
		P	S	D
Ta et al. (2009)	Availability of more than one resource to provide a system function	X		
Rice and Caniato (2003)	Maintain capacity to respond to disruptions in the supply network	X	X	
Christopher and Peck (2004)	Keep several options open despite possible increased costs	X	X	
Sheffi and Rice (2005)	To keep some resources in reserve to be used in case of a disruption	X	X	
Albert and Barabási (2002)	Existence of many alternative paths that can preserve communication between nodes even if some nodes are absent			X
Cholda et al. (2007)	Additional resource usage necessary to support a selected recovery method			X
Sterbenz et al. (2011)	Replication of entities in the network, generally to provide fault-tolerance			X
Jackson and Ferris (2013)	Two or more different ways to perform a critical task	X	X	X

[a]P physical, S service, D digital

levels of resilience, agent-level redundancy aims at providing local resilience while network-level redundancy can enable global resilience. Strategies for redundancy at local and global levels are discussed in the next sections.

4.2.1 Agent-Level (or Local) Redundancy

Over the years, several authors have developed methodologies to create agent-level redundancy in order to ensure a SN agent can achieve a desired QoS even when its predecessors and/or successors experience disruptions that restrict flow. In digital SNs, local redundancy relates to the existence of multiple links between a SN agent a and other neighboring SN agents to enable fault tolerant flow of digital information (Jeong 2009; Erdene-Ochir et al. 2010; Sterbenz et al. 2011). Redundancy must be enabled and exploited within limited energy availability; therefore, topology control protocols must optimize the location and allocation of redundancy by dynamically creating and eliminating flow links.

Santi (2005) and Li et al. (2013) review topology control protocols for WSNs and discuss fault-tolerant alternatives for topology control based on transmission power control with full or partial agent location information. Kim et al. (2009) develop an algorithm to detect critical agents with low number of connections in WSNs topologies and analyze the effect of two protection strategies: (1) creating a full mesh around critical and under-connected agents, (2) creating sufficient connections to eliminate agent criticality. Numerical tests show that smart creation of additional connections to reduce/eliminate node criticality while minimizing the cost of the new connections leads to more resilient networks than when imposing a minimum number of connections as a protective measure. Similarly, Jiang et al. (2014) analyze the effect of re-configuration strategies based on node degree and betweenness centrality on a digital SN's capacity to deliver flow. Numerical results show that intentional re-configuration to achieve a more homogeneous betweenness centrality can augment the SN's efficiency to distribute flow, enhancing its ability to cope with congestion.

In general, each flow control protocol in digital SNs will require specific topological characteristics, some of which may be related to local redundancy, in order to guarantee the desired level resilience. For instance, in sensor networks, the fault-tolerant time-out protocol (FTTP) developed by Liu and Nof (2008) requires that sensors are arranged following a cluster network architecture. This arrangement entails forming communities (clusters) of sensors with a designated cluster head. Each community consists on a small number of sensors that are redundantly interconnected, where the cluster head is responsible for executing the FTTP to obtain readings from all sensors. Communities are, in turn, arranged following the cluster network architecture around a base station, the output agent of the digital SN. To minimize energy consumption while operating under FTTP, Jeong and Nof (2008, 2009) develop a formation and re-configuration algorithm to dynamically control clusters' topology.

In physical supply networks, the strategy of local redundancy has mainly focused on the use of multiple predecessors, normally termed suppliers, under different sourcing protocols to ensure sufficient input flow in order to maintain desired QoS (e.g., Minner 2003; Rice and Caniato 2003; Tang 2006; Federgruen and Yang 2008; Tang and Tomlin 2008; Pettit et al. 2010; Spiegler et al. 2012; Silbermayr and Minner 2014). Supplier selection has been extensively addressed in SN literature—see Weber et al. (1991) and Aissaoui et al. (2007) for a comprehensive review of selection approaches; nevertheless, resilience requirements have not been considered until recently by Sawik (2013). The author combines mathematical optimization with value-at-risk metrics to derive optimal portfolios of suppliers in terms of cost and exposure to disruption risk.

Less frequently, authors addressed the need for successor redundancy in order to cope with disruptions in downstream demand; for instance, Tang and Tomlin (2008) use successor redundancy in combination with responsive pricing protocols to shift demand across successors in the event of demand disruptions. Link redundancy is also required when the focus is not to ensure that SN agent *a* maintains overall QoS but rather that SN agent *a* delivers pre-arranged QoS to a specific successor. In this line, Pettit et al. (2010) propose the use of alternative distribution channels in order to have flexibility in order fulfillment and (Rice and Caniato 2003) suggest the use of multiple transportation modes and carriers.

Local redundancy in physical SNs also encompasses lateral collaboration, i.e., the formation of links between SN agents that can produce the same type of output flow. Lateral collaboration enables (1) production capacity sharing to meet individual agents' output demand (Yoon and Nof 2010, 2011; Seok and Nof 2014a, b; Moghaddam and Nof 2014) and (2) inventory sharing, normally termed lateral transshipments, to cope with demand and supply fluctuations and disruptions (Minner 2003; Paterson et al. 2011; Smirnov and Gerchak 2014). It can be argued that lateral collaboration may extend to supplier sharing, e.g., sharing of surplus supplier capacity and reallocation of supply orders (SN agent i allocates one of its suppliers' future deliveries to SN agent j to help the latter overcome one of its own suppliers' disruptions).

When designing for local redundancy, it is important to consider potential correlation among redundant predecessors/successors, a condition that has been termed *fate sharing* (Craighead et al. 2007; Sterbenz et al. 2011). Fate sharing emerges when two or more SN agents or links share a common characteristic that may cause them to be disrupted simultaneously, e.g., geographical proximity (Tang 2006; Craighead et al. 2007; Falasca et al. 2008; Ratick et al. 2008; Sterbenz et al. 2011), common predecessors, and shared inherent design flaws.

4.2.2 Network-Level (or Global) Redundancy

From a global perspective, local redundancy increases the number of interconnections and interactions. Craighead et al. (2007) argue that a high level of interconnection can be detrimental to SN resilience, as it helps propagate the effects of local disruptions to more SN agents. On the other hand, it can also help increase resilience by providing alternative paths to overcome faulty nodes/links (Albert and Barabási 2002; Falasca et al. 2008).

Network-level (or global) redundancy consists on creating multiple paths to connect any pair of source and sink agents in a SN or, in general, to connect any two SN agents (Rohrer et al. 2009; Ta et al. 2009; Erdene-Ochir et al. 2010; Sterbenz et al. 2011). Although local redundancy can help increase individual SN agent's resilience, it does not guarantee network-level resilience unless local redundancy is correctly allocated among SN agents. For instance, consider the alternative configurations of a SN shown in Fig. 4.1. Both configurations have agents with local redundancy and equal global redundancy in terms of number of paths (i.e., 3) between dark shaded agents. Nevertheless, all paths in SN I contain link A and are therefore simultaneously affected by disruptions to the common link whereas in SN II no single-link disruption disables all paths at the same time.

Regarding the allocation of local redundancy, some authors e.g., Hearnshaw and Wilson (2013) argue that a SN must have a power-law degree distribution (Barabási and Albert 1999), in which a small fraction of agents are highly connected while the majority present a low degree (no. of connections). This provides tolerance to random disruptions, i.e., affecting any agent/link with equal probability, but is not able to cope with targeted disruptions, i.e., those purposely affecting highly connected nodes (Albert et al. 2000).

Fig. 4.1 Alternative SN
topologies with different
resilience to link disruption

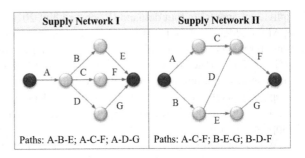

Recently, Brede and de Vries (2009) have suggested that assortative mixing is important to provide enhanced resilience to failure of highly connected agents. In SNs with high assortative mixing, agents with high degree will tend to form a sub-network (or community) by connecting to each other; a core–periphery topology arises where core (or kernel) agents have similar degrees and are strongly interlinked with each other—this is sometimes called the rich-club phenomenon (Colizza et al. 2006). This configuration among kernel agents may provide an added layer of resilience to targeted disruptions. A similar phenomenon is observed by Reyes Levalle and Nof (2015) in a network where fault-prone agents collaborate to form teams in order to protect from, and overcome, disruptions. Agents which have an incentive to form teams modify an otherwise scale-free structure by creating more interconnections among moderately connected agents, creating the conditions for teaming protocols to leverage these redundant paths and adequately re-route flow in the occurrence of a disruption.

Another important global redundancy property, especially in settings where redundancy through lateral collaboration is designed at local level, is the clustering coefficient which describes the extent to which any three agents in a SN are interconnected (Watts and Strogatz 1998). A high clustering coefficient favors short paths between any two agents, a phenomenon known as small-worldness. Nevertheless, the effect of this topological configuration on lateral collaboration and its effect on global resilience have not been addressed in literature hitherto.

Redundancy is mainly concerned with agent-level and network-level connectivity to enable multiple alternative flow paths between agents. Nevertheless, several authors argue that network-level redundancy must also encompass the use of distributed (excess) resources and capabilities. In physical SNs, Rice and Caniato (2003), Sheffi and Rice (2005), Tang and Tomlin (2008), and Pettit et al. (2010) suggest that production facilities must be designed with backup and interoperability, i.e., there must be at least two SN agents with equal capabilities; distributed capacity and capability enables the SN to shift production demands among agents to accommodate for demand variations and disruptions (Tang and Tomlin 2008).

Following Jackson and Ferris (2013), the above discussed concept can be formalized in a general design principle: *SNs functionality must be distributed among various agents so that in the event a disruption affects a subset of the SN agents, the remaining ones are able to maintain QoS within the SLA limits*. In order to design a

SN using this principle, it is necessary to combine redundancy with a second resilience strategy, namely, excess resources, which is discussed in the next section.

4.3   Excess Resources: Storage and Capacity Surplus

In addition to the use of redundancy, SN designers can strategically locate excess resources throughout the network to protect key components from the occurrence of disruptions and to minimize the time to recover from a disruptive event. Storage provides protection against phenomena known as *blocking* and *starvation* in production lines literature. In other words, having additional storage capacity can help maintain flows when a SN agent is disrupted by supplying its downstream SN agents after a disruption and/or allowing the SN agent to maintain operations when its successors are inoperable and cannot receive flow.

Capacity surplus, in combination with network-level redundancy, protects network-level flows from disruptions by enabling alternate flow paths to supply downstream SN agents. Furthermore, capacity surplus aids the recovery process by allowing transient increases in local flow to stabilize operations in the vicinity of a recently disrupted agent.

4.3.1   Capacity Surplus

Rice and Caniato (2003), Christopher and Peck (2004), Sheffi and Rice (2005), and Pettit et al. (2010) call for capacity surplus in physical SNs as a way of coping with disruptions. At a local level, capacity surplus can reduce the effect of a disruption by shortening the required recovery time or even prevent an error/conflict from becoming a disruption by activating parallel unused resources. For instance, a production line designed with a production rate higher than the average required to fulfill the SLA with successors is able to use the excess rate after a disruption to recover faster than in the case of tight capacity. A more efficient approach to design of capacity surplus is to parallelize the available excess capacity; if errors/conflicts are uncorrelated, parallelism enables the system to retain most of its functionality through a larger fraction of its operational time, thus reducing, or even eliminating, the occurrence of disruptions (and reducing their effect in case of occurrence).

The concept of surplus capacity as a resilience enabler also applies to digital SNs. Menth et al. (2006) analyze the use of capacity overprovisioning in packet-switched communication networks to overcome traffic variations due to demand variability and disruptions. Cholda et al. (2007) review frameworks for resilient communication networks and categorize the use of spare capacity in dedicated and shared, depending on whether it is assigned to a single SN agent or used for shared recovery among various SN agents. The existence of alternatives allows SN agents to apply resilience differentiation protocols to provide successors with alternative levels of resilience, at different cost. Sterbenz et al. (2011) highlight the importance of selecting the right mix of capacity surplus among different SN resources, e.g., bandwidth, processor capacity, and latency, and the optimization of surplus allocation among SN agents to achieve the most cost-efficient SN design.

In both digital and physical SNs, authors point out the trade-off between increased number of resources available and cost (Christopher and Peck 2004; Sterbenz et al. 2011). In order to minimize the investment and operational burden associated to capacity surplus, network design for operation under capacity sharing protocols (Yoon and Nof 2010, 2011; Seok and Nof 2014a, b; Moghaddam and Nof 2014) that dynamically optimize the match between available capacity and resilience needs appear as a viable solution, but needs further exploration.

4.3.2   Storage

The strategy of protection against disruptions by holding inventory is often referred to among design for resilience authors. For instance, Jackson and Ferris (2013) introduce the principle of absorption by which an engineered system should be capable of absorbing the magnitude of a disruption. Although inventory can provide this capability, it can lead to significant costs associated to inventory holding and longer leadtimes, making the strategy effective only against short and frequent disruptions that require low inventory levels (Tomlin 2006).

Formally, the capability designed into a SN that enables agents to hold inventory is storage; it involves excess resources such as space and storage equipment, and needs handling protocols to efficiently manage these excess resources. Storage creates loose-coupling or even decoupling among SN agents so that disruptions do not disseminate or are quickly contained (Hearnshaw and Wilson 2013; Jackson and Ferris 2013). Mathematical proofs from queuing network theory show that, ceteris paribus, larger storage capacity (buffer size) leads to higher throughput (Cruz et al. 2010; Smith et al. 2010; Smith 2014).

Designing storage capabilities in a SN is a location-allocation problem; it involves defining which SN agents must be assigned storage capabilities and how much storage to assign to each agent. Gershwin and Schor (2000) and Demir et al. (2012) review buffer allocation strategies for production systems, in which an available amount of storage must be optimally allocated to production resources in order to accomplish a specific objective, e.g., maximizing throughput or minimizing work-in-process (WIP). Storage reduces idle time due to starvation (lack of input flow from predecessors) and blocking (impossibility to deliver output flow to successors) due to upstream and downstream disruptions. Despite the number of available analyses on buffer allocation, most focus on production lines, not SNs, and only two consider simultaneously WIP minimization and throughput maximization, and none address minimization of throughput variability; all three conditions required for resilience.

The aforementioned reviews of storage location-allocation problems show that the design has been addressed by centralized optimization models. An emerging possibility is to investigate decentralized location-allocation models based on centrality and/or criticality measures.

4.4   Communication Network Efficiency

One of the fundamentals of resilience is the ability of agents and SNs to anticipate disruptions and implement (1) preventive actions in order to eliminate the forthcoming disruption or (2) preparedness actions to eliminate, or at least minimize, the impact thereof. To this end, it is essential to design a communication network $CN = (A, CL)$ for efficient information flow among SN agents and to embed situation awareness capabilities in SN agents.

Situation awareness is the first component required to anticipate and avoid or prepare for disruptions. In order for advance information regarding an error/conflict possibly leading to a disruption to be effectively used, it must be efficiently disseminated throughout the SN to those agents that can make collaborative decisions and implement for avoidance/preparedness actions based on advance information (Craighead et al. 2007). As defined by the principle of emergent lines of collaboration and command, LOCC, from Collaborative Control Theory (Nof 2007; Velásquez and Nof 2009) efficient, cost-effective information delivery for collaborative decision-making requires to clearly define which SN agents to contact, under what conditions to establish contact, when to initiate contact, and how to begin contact under time pressure.

Communication between any two agents can be direct or indirect, i.e., via a third or more agents; therefore, contact requirements for each SN agent emerging from the LOCC principle can be met by several alternative communication network topologies. Unfortunately, there is a dearth of research on optimal topological properties for communication networks in SNs. Christopher and Peck (2004) discuss the need for communities of SN agents to share information arising from situation awareness to reduce uncertainty and enable risk management; however, Danon et al. (2008) conclude that strong community structures in SNs can create bottlenecks that affect timely information dissemination.

Watts and Strogatz (1998) developed a network formation mechanism that gives rise to highly clustered networks with short characteristic path length (CPL), named small-world networks. The characteristic path length of a network (Watts and Strogatz 1998) is a measure of the average distance between any pair of agents. Watts and Strogatz (1998) conclude that small-world networks enable faster propagation speed for signals and infectious diseases, when compared to random and regular networks. Following this observation, Hearnshaw and Wilson (2013) suggest that a short characteristic path length is required for information to diffuse and circulate information rapidly throughout the SN. Natural design seems to be in line with this observation; Rubinov and Sporns (2010) review studies examining brain networks and observe that in several cases such networks are organized as small-world networks.

Despite the emerging work in the design and analysis of communication networks to support situation awareness and efficient communication for resilience, further research efforts are required to more deeply understand the implications and needs of this strategy for resilience in supply networks.

4.5 Summary and Outlook

In this chapter, the three main strategies to design resilient supply networks—redundancy, excess resources, communication network efficiency—were introduced and discussed in connection with several existing techniques found in digital, physical, and service supply networks. The mapping developed in the above sections provides practitioners and researchers a common framework to analyze and design existing and new approaches to designing resilient structures in SNs.

Hitherto, SN research has mainly addressed redundancy and excess resources strategies to enable resilience. Strategies to design efficient and effective communication networks for resilience in SNs are still lacking in current literature. Further analysis is needed to identify the main characteristics a communication network $CN = (A, CL)$ requires to enable adequate interaction for resilience. Insights gained from these efforts will support ongoing design and re-design of organizations and systems, to support resilient operations.

Lastly, a note on a trend found, mostly, in physical SNs. Efforts have focused typically on understanding the conditions required to enable local resilience, i.e., at agent level. Myopic application of agent-level resilience strategies may not necessarily lead to network-level resilience and, in some cases, it may even be detrimental to this end. Further research is needed to understand the network-level implications of applying local strategies; such efforts are emergent and hinge on the rise of network analysis and increase computational capacity which enables larger agent-based models to re-create real-world systems and their dynamics.

References

Aissaoui, N., Haouari, M., & Hassini, E. (2007). Supplier selection and order lot sizing modeling: A review. *Computers & Operations Research, 34*(12), 3516–3540. doi:10.1016/j.cor.2006.01.016

Al-Karaki, J. N., & Kamal, A. E. (2004). Routing techniques in wireless sensor networks: A survey. *IEEE Wireless Communications, 11*(6), 6–28. doi:10.1109/MWC.2004.1368893

Albert, R., & Barabási, A.-L. (2002). Statistical mechanics of complex networks. *Reviews of Modern Physics, 74*(1), 47–97. doi:10.1103/RevModPhys.74.47

Albert, R., Jeong, H., & Barabási, A.-L. (2000). Error and attack tolerance of complex networks. *Nature, 406*(6794), 378–382. doi:10.1038/35019019

Barabási, A.-L., & Albert, R. (1999). Emergence of scaling in random networks. *Science, 286* (5439), 509–512. doi:10.1126/science.286.5439.509

Brede, M., & de Vries, B. J. M. (2009). Networks that optimize a trade-off between efficiency and dynamical resilience. *Physics Letters A, 373*(43), 3910–3914. doi:10.1016/j.physleta.2009. 08.049

Cholda, P., Mykkeltveit, A., Helvik, B., Wittner, O., & Jajszczyk, A. (2007). A survey of resilience differentiation frameworks in communication networks. *IEEE Communications Surveys & Tutorials, 9*(4), 32–55. doi:10.1109/COMST.2007.4444749

Christopher, M., & Peck, H. (2004). Building the resilient supply chain. *International Journal of Logistics Management, 15*(2), 1–14. doi:10.1108/09574090410700275

Colizza, V., Flammini, A., Serrano, M. A., & Vespignani, A. (2006). Detecting rich-club ordering in complex networks. *Nature Physics, 2*(2), 110–115. doi:10.1038/nphys209

Craighead, C. W., Blackhurst, J., Rungtusanatham, M. J., & Handfield, R. B. (2007). The severity of supply chain disruptions: Design characteristics and mitigation capabilities. *Decision Sciences, 38*(1), 131–156. doi:10.1111/j.1540-5915.2007.00151.x

Cruz, F. R. B., Van Woensel, T., & Smith, J. M. (2010). Buffer and throughput trade-offs in M/G/1/K queueing networks: A bi-criteria approach. *International Journal of Production Economics, 125*(2), 224–234. doi:10.1016/j.ijpe.2010.02.017

Danon, L., Arenas, A., & Díaz-Guilera, A. (2008). Impact of community structure on information transfer. *Physical Review E, 77*(3), 36103. doi:10.1103/PhysRevE.77.036103

Demir, L., Tunali, S., & Eliiyi, D. T. (2012). The state of the art on buffer allocation problem: A comprehensive survey. *Journal of Intelligent Manufacturing, 25*(3), 371–392. doi:10.1007/ s10845-012-0687-9

Erdene-Ochir, O., Minier, M., Valois, F., & Kountouris, A. (2010). Resiliency of wireless sensor networks: Definitions and analyses. In *2010 IEEE 17th International Conference on Telecommunications* (pp. 828–835). Doha, Qatar. doi:10.1109/ICTEL.2010.5478822

Falasca, M., Zobel, C. W., & Cook, D. (2008). A decision support framework to assess supply chain resilience. In: *Proceedings of the 5th International ISCRAM Conference* (pp. 596–605). Washington, DC, USA.

Federgruen, A., & Yang, N. (2008). Selecting a portfolio of suppliers under demand and supply risks. *Operations Research, 56*(4), 916–936. doi:10.1287/opre.1080.0551

Gershwin, S. B., & Schor, J. E. (2000). Efficient algorithms for buffer space allocation. *Annals of Operations Research, 93*(1–4), 117–144. doi:10.1023/A:1018988226612

Gong, J., Mitchell, J. E., Krishnamurthy, A., & Wallace, W. A. (2014). An interdependent layered network model for a resilient supply chain. *Omega, 46,* 104–116. doi:10.1016/j.omega.2013. 08.002

Hearnshaw, E. J. S., & Wilson, M. M. J. (2013). A complex network approach to supply chain network theory. *International Journal of Operations & Production Management, 33*(4), 442–469. doi:10.1108/01443571311307343

Jackson, S., & Ferris, T. L. J. (2013). Resilience principles for engineered systems. *Systems Engineering, 16*(2), 152–164. doi:10.1002/sys.21228

Jeong, W. (2009). Sensor and sensor networks. In S. Y. Nof (Ed.), *Springer handbook of automation* (pp. 333–348). Berlin: Springer. doi:10.1007/978-3-540-78831-7_20

Jeong, W., & Nof, S. Y. (2009). A collaborative sensor network middleware for automated production systems. *Computers & Industrial Engineering, 57*(1), 106–113. doi:10.1016/j.cie. 2008.11.007

Jeong, W., & Nof, S. Y. (2008). Performance evaluation of wireless sensor network protocols for industrial applications. *Journal of Intelligent Manufacturing, 19*(3), 335–345. doi:10.1007/ s10845-008-0086-4

Jiang, Z.-Y., Liang, M.-G., & An, W.-J. (2014). Effects of efficient edge rewiring strategies on network transport efficiency. *Physica A: Statistical Mechanics and its Applications, 394,* 379–385. doi:10.1016/j.physa.2013.09.069

Kim, T., Tipper, D., Krishnamurthy, P., & Swindlehurst, A. L. (2009). Improving the topological resilience of mobile ad hoc networks. In *Proceedings of the 7th International Workshop on Design of Reliable Communication Networks* (pp. 191–197). Washington, DC, USA. doi:10.1109/DRCN.2009.5340008

Li, M., Li, Z., & Vasilakos, A. V. (2013). A survey on topology control in wireless sensor networks: Taxonomy, comparative study, and open issues. *Proceedings of the IEEE, 101*(12), 2538–2557. doi:10.1109/JPROC.2013.2257631

Liu, Y., & Nof, S. Y. (2008). Fault-tolerant sensor integration for micro flow-sensor arrays and networks. *Computers & Industrial Engineering, 54*(3), 634–647. doi:10.1016/j.cie.2007.09.013

Menth, M., Martin, R., & Charzinski, J. (2006). Capacity overprovisioning for networks with resilience requirements. *ACM SIGCOMM Computer Communication Review, 36*(4), 87. doi:10.1145/1151659.1159925

Minner, S. (2003). Multiple-supplier inventory models in supply chain management: A review. *International Journal of Production Economics, 81–82,* 265–279. doi:10.1016/S0925-5273(02)00288-8

Moghaddam, M., & Nof, S. Y. (2014). Combined demand and capacity sharing with best matching decisions in enterprise collaboration. *International Journal of Production Economics, 148,* 93–109. doi:10.1016/j.ijpe.2013.11.015

Nof, S. Y. (2007). Collaborative control theory for e-Work, e-Production, and e-Service. *Annual Reviews in Control, 31*(2), 281–292. doi:10.1016/j.arcontrol.2007.08.002

Paterson, C., Kiesmüller, G., Teunter, R., & Glazebrook, K. (2011). Inventory models with lateral transshipments: A review. *European Journal of Operational Research, 210*(2), 125–136. doi:10.1016/j.ejor.2010.05.048

Pettit, T. J., Fiksel, J., & Croxton, K. L. (2010). Ensuring supply chain resilience: Development of a conceptual framework. *Journal of Business Logistics, 31*(1), 1–21. doi:10.1002/j.2158-1592.2010.tb00125.x

Ratick, S., Meacham, B., & Aoyama, Y. (2008). Locating backup facilities to enhance supply chain disaster resilience. *Growth and Change, 39*(4), 642–666. doi:10.1111/j.1468-2257.2008.00450.x

Reyes Levalle, R., & Nof, S. Y. (2015). Resilience by teaming in supply network formation and re-configuration. *International Journal of Production Economics, 160,* 80–93. doi:10.1016/j.ijpe.2014.09.036

Rice, J. B., Jr., & Caniato, F. (2003). Building a secure and resilient supply network. *Supply Chain Management Review, 7*(5), 22–30.

Rohrer, J. P., Jabbar, A., & Sterbenz, J. P. G. (2009). Path diversification: A multipath resilience mechanism. In *Proceedings of the 7th International Workshop on Design of Reliable Communication Networks* (pp. 343–351). Washington, DC, USA. doi:10.1109/DRCN.2009.5339988

Rubinov, M., & Sporns, O. (2010). Complex network measures of brain connectivity: Uses and interpretations. *NeuroImage, 52*(3), 1059–1069. doi:10.1016/j.neuroimage.2009.10.003

Santi, P. (2005). Topology control in wireless ad hoc and sensor networks. *ACM Computing Surveys, 37*(2), 164–194. doi:10.1145/1089733.1089736

Sawik, T. (2013). Selection of resilient supply portfolio under disruption risks. *Omega, 41*(2), 259–269. doi:10.1016/j.omega.2012.05.003

Seok, H., & Nof, S. Y. (2014a). Collaborative capacity sharing among manufacturers on the same supply network horizontal layer for sustainable and balanced returns. *International Journal of Production Research, 52*(6), 1622–1643. doi:10.1080/00207543.2013.842016

Seok, H., & Nof, S. Y. (2014b). Dynamic coalition reformation for adaptive demand and capacity sharing. *International Journal of Production Economics, 147,* 136–146. doi:10.1016/j.ijpe.2013.09.003

Sheffi, Y., & Rice, J. B., Jr. (2005). A supply chain view of the resilient enterprise. *MIT Sloan Management Review, 47*(1), 41–48.

Silbermayr, L., & Minner, S. (2014). A multiple sourcing inventory model under disruption risk. *International Journal of Production Economics, 149,* 37–46. doi:10.1016/j.ijpe.2013.03.025

Smirnov, D., & Gerchak, Y. (2014). Inventory sharing via circular unidirectional chaining. *European Journal of Operational Research, 237*(2), 474–486. doi:10.1016/j.ejor.2014.02.019

Smith, J. M. (2014). System capacity and performance modelling of finite buffer queueing networks. *International Journal of Production Research, 52*(11), 3125–3163. doi:10.1080/00207543.2013.854935

Smith, J. M., Cruz, F. R. B., & van Woensel, T. (2010). Topological network design of general, finite, multi-server queueing networks. *European Journal of Operational Research, 201*(2), 427–441. doi:10.1016/j.ejor.2009.03.012

Spiegler, V. L. M., Naim, M. M., & Wikner, J. (2012). A control engineering approach to the assessment of supply chain resilience. *International Journal of Production Research, 50*(21), 6162–6187. doi:10.1080/00207543.2012.710764

Sterbenz, J. P. G., Cetinkaya, E. K., Hameed, M. A., Jabbar, A., Qian, S., & Rohrer, J. P. (2011). Evaluation of network resilience, survivability, and disruption tolerance: Analysis, topology generation, simulation, and experimentation. *Telecommunication Systems, 52*(2), 705–736. doi:10.1007/s11235-011-9573-6

Ta, C., Goodchild, A. V., & Pitera, K. (2009). Structuring a definition of resilience for the freight transportation system. *Transportation Research Record: Journal of the Transportation Research Board, 2097,* 19–25. doi:10.3141/2097-03

Tang, C. S. (2006). Perspectives in supply chain risk management. *International Journal of Production Economics, 103*(2), 451–488. doi:10.1016/j.ijpe.2005.12.006

Tang, C. S., & Tomlin, B. (2008). The power of flexibility for mitigating supply chain risks. *International Journal of Production Economics, 116*(1), 12–27. doi:10.1016/j.ijpe.2008.07.008

Tomlin, B. (2006). On the value of mitigation and contingency strategies for managing supply chain disruption risks. *Management Science, 52*(5), 639–657. doi:10.1287/mnsc.1060.0515

Velásquez, J. D., & Nof, S. Y. (2009). Collaborative e-work, e-business, and e-service. In S. Y. Nof (Ed.), *Springer handbook of automation* (pp. 1549–1576). Berlin: Springer. doi:10.1007/978-3-540-78831-7_88

Watts, D. J., & Strogatz, S. H. (1998). Collective dynamics of 'small-world' networks. *Nature, 393*(6684), 440–442. doi:10.1038/30918

Weber, C. A., Current, J. R., & Benton, W. C. C. (1991). Vendor selection criteria and methods. *European Journal of Operational Research, 50*(1), 2–18. doi:10.1016/0377-2217(91)90033-R

Yoon, S. W., & Nof, S. Y. (2011). Affiliation/dissociation decision models in demand and capacity sharing collaborative network. *International Journal of Production Economics, 130*(2), 135–143. doi:10.1016/j.ijpe.2010.10.002

Yoon, S. W., & Nof, S. Y. (2010). Demand and capacity sharing decisions and protocols in a collaborative network of enterprises. *Decision Support Systems, 49*(4), 442–450. doi:10.1016/j.dss.2010.05.005

Chapter 5
Flow Control Protocols for Resilient Supply Networks

Resilience is a dynamic property of SNs. The ability of a SN to minimize the effect of disruptions on its performance depends on the interaction between its structural properties and the set of flow control protocols used to dynamically assign and re-assign flow and resources. Hence, SN design should not only be concerned with the structural or control aspects in isolation but rather with the integrated design of these and their seamless interaction.

Flow control protocols manage digital, physical, and service flows, and communication exchanges, among and within SN agents in order to conduct operations under normal and disrupted conditions, without altering existing SN structure. In this chapter, the three main types of flow control in SNs—sourcing control, internal resource control, and distribution control—are characterized and their interrelation with SN structural characteristics is discussed within the SN formalism introduced in Chap. 2. Several examples from literature illustrate applications that belong to each of the three protocol types.

5.1 Sourcing Protocols: Controlling Input Flow

Following the SN formalism presented in Chap. 2, kernel and sink agents $i \in A^K \cup A^O$ need to obtain flow from their predecessors $j \in P_i \subseteq A^I \cup A^K$ in order to fulfill their SLAs with their successors $h \in S_i \subseteq A^K \cup A^O$. Through sourcing control protocols, a SN agent i interacts with its predecessors P_i to decide on (1) the characteristics of the required flow and (2) the subset of agents $j \in P_i$ that will supply the required flow.

From the requesting agent's perspective, the process from issuance to completion of a sourcing request comprises three activities: (1) waiting and (2) processing, both at agent j, and (3) delivery to agent i. Waiting before processing depends on agent j's demand, affected by the number and characteristics of requests made by its

© Springer International Publishing AG 2018
R. Reyes Levalle, *Resilience by Teaming in Supply Chains and Networks*,
Automation, Collaboration, & E-Services 5, DOI 10.1007/978-3-319-58323-5_5

successors S_j, as well as its own processing capacity. Processing time depends strictly on agent j's capacity and agent i's sourcing request characteristics, and delivery depends on link $fl_{i \rightarrow j}$ conditions. The total time between the sourcing request's issuance and completion is generally termed *leadtime*.

Leadtime variability, a phenomenon that can lead to disruptions in a SN agent making a sourcing request, has been subject of extensive research in physical SNs. Variability may be the result of inherent randomness in the sourcing process, e.g., travel time randomness due to route conditions, and/or disruptions in upstream processes, e.g., machine breakdowns in predecessors. Notwithstanding the cause of variability, sourcing protocols must be able to leverage network structural characteristics to avoid disruptions at the requesting agent.

Sourcing control protocols to operate with resilience under leadtime variability can be grouped in three categories: multi-sourcing, back-up sourcing, and emergency sourcing. In multi-sourcing, agent i requests flow simultaneously from a subset of its predecessors P_i to fulfill its needs. The strategy requires structural redundancy and storage capacity, and, oftentimes, real-time request status information exchange. Back-up sourcing differs from multi-sourcing in that there is a layered approach to requesting flow from predecessors; initial requests are submitted to primary sourcing agents, a subset of P_i and only in the event that some/all of the primary agents fail to deliver the requested flow it is re-assigned to back-up agents. The structural requirements for back-up sourcing coincide with those of multi-sourcing. Emergency sourcing differs from the previous strategies in that agent i leverages the excess capacity of a predecessor to transitorily, and, usually, with short a notice, increase sourcing flow. Table 5.1 summarizes the structural requirements of each sourcing protocol category.

5.1.1 Multi-sourcing

In multi-sourcing, each request made by SN agent i comprises a subset of requirements, each of which must be fulfilled by a different predecessor $j \in P_i$. In most cases, all predecessors are expected to supply the requested flow; nevertheless, there is some tolerance in terms of when they are required to fulfill their supply agreements. Multi-sourcing occurs naturally when no single agent j can supply agent i's requests or can be purposely pursued by agent i to diversify the risk of disruptions and their effects.

In digital SNs, multi-sourcing is commonly found, in particular in sensor networks. The use of clusters of low cost sensors to obtain replicated data is increasingly used as a means to overcome faults to single sensors/links and/or increase readings' accuracy. Among communication protocols in sensor networks, two protocols based on the Fault Tolerance by Teaming principle of Collaborative Control Theory (Nof 2003) are of particular interest to resilient sourcing:

Table 5.1 Structural requirements for resilient sourcing protocols

Category	Description	SN structure requirements			
		Redundancy	ER[a]-storage	ER[a]-Cap. surplus	SA + ECN[a]
Multi-sourcing	Sourcing requests are split among several predecessors; most/all parts are needed to complete the request	R	R	NR	PD
Back-up sourcing	Sourcing requests are assigned to a set of primary predecessors; if one/several are not available, requests are sent to back-up predecessors	R	R	NR	PD
Emergency sourcing	A predecessor with excess capacity is assigned a sourcing request; if needed, leadtime can be shortened (at a cost)	NR	R	R	PD

[a]*ER* excess resources, *ECN* efficient communication network, *SA* situation awareness, *R* required, *NR* not required, *PD* protocol-dependent

fault-tolerant time-out protocol, FTTP, and fault-tolerant sensor integration algorithm, FTSIA.

In FTTP, Liu and Nof (2004) leverage the redundant nature of a clustered sensor architecture to overcome faulty links; the protocol issues a sourcing request to all predecessors of a base station and, in the event one predecessor's reading is not received after a predefined time, the missing data is requested from a sibling node. Numerical experiments show that, when operating under FTTP, sourcing requests can be completed even after 50% the network links are eliminated.

FTSIA (Liu and Nof 2008) is designed to enable the sensor cluster to overcome faulty sensor readings and increase combined sensor accuracy. The algorithm uses a statistical-driven logic to identify faulty and possibly faulty readings, and determine which subset of the predecessors' data is to be combined into a given measure. Experiments show that FTSIA is able to provide more reliable and accurate readings when faulty readings are present than other widely accepted techniques commonly used in industry.

The aforementioned protocols, FTTP and FTSIA, provide a valuable insight in the context of resilience: performance of fault-prone agents can be augmented by collaboration and teaming, to the extent that combined performance can exceed that of a single flawless agent; this is the essence of the Fault-tolerance by Teaming principle of Collaborative Control Theory (Nof 2007).

In physical SNs, the research landscape presents mixed results. Several authors argue that there are benefits to multi-sourcing, which can be distilled in three theoretical advantages: (1) effective leadtime, i.e., the arrival time of the first delivery, is always shorter than the average leadtime of any of the predecessors used as a single source to complete a request (Pan et al. 1991), (2) fully reliable sourcing is asymptotically achievable with a sufficiently large number of predecessors (Federgruen and Yang 2008), (3) lower average inventory (Hill 1996). These benefits can be further extended if deliveries from predecessors are scheduled to be received sequentially at the requesting agent (Glock and Ries 2013).

Despite the observed benefits, some recent work suggest that the above results may rely heavily on strong assumptions that may not hold in real-world SNs. Thomas and Tyworth (2006) re-evaluate the benefits of multi-sourcing models from a network perspective and conclude that, in previous analysis, benefits from cycle inventory reduction have been overestimated by not considering pipeline inventory. Furthermore, any costs reductions emerging from lower inventory holding costs could be outweighed by including transportation costs, oftentimes assumed invariant with respect to the number of deliveries.

5.1.2 Back-up Sourcing

In a back-up sourcing setting, a SN agent i establishes a priority ranking among its predecessors P_i according to some predefined criteria and assigns sourcing requests to the top ranked available agent $j \in P_i$. In the case an agent j becomes unable to fulfill a sourcing request, the latter is reassigned to the top ranked available agent $\in P_i$, $k \neq j$. Normally, whenever agent i reassigns a sourcing request to a back-up predecessor, it incurs in some form of excess cost emerging from, e.g., the need of shorter leadtime, delivery guarantees, or spot market conditions, in physical SNs or increased energy consumption due to repeated transmissions and processing in digital SNs.

A commonly found implementation of back-up sourcing in physical SNs consists on ranking agents in P_i based on decreasing cost; assuming Pareto efficiency, this ranking method implies that agents are also ranked according to increasing reliability. Tomlin (2006) proposes the use of option contracts to implement back-up sourcing. In this setting, agent i reserves surplus capacity at a subset of P_i by paying a reservation prime in advance; thus, agent i has the right to exercise the option to receive extra flow when needed. Analyses in a two predecessor scenario show that back-up sourcing becomes less attractive relative to regularly sourcing from a more reliable predecessor as the frequency and/or duration of unreliable predecessors' disruptions increases.

Asian and Nie (2014) expand traditional analysis of option contracts by modeling back-up agents' speculation. The authors analyze coordination mechanisms when the back-up source will reserve a fraction of its capacity that may be lower than that agreed in an option contract. Such behavior enables the back-up agent to increase its own benefits but exposes its successors to higher risk—even under the option contract. The use of option contracts can be optimized by using a portfolio approach, as suggested by Fu et al. (2010). Nevertheless, the mechanism for increasing resilience in back-up sourcing will always rely on the existence of back-up agents with the capacity to outperform the primary predecessor; it is not possible to leverage the performance of weaker agents as in the case of multi-sourcing protocols.

Demand and capacity sharing (DCS) protocols, developed to cope with inherent demand variability by providing agents with the opportunity to leverage unused capacity and otherwise lost demand to achieve mutual benefits (Yoon and Nof 2010; Moghaddam and Nof 2014; Seok and Nof 2014), belong to the category of back-up sourcing. In DCS, an agent i unable to fulfill a sourcing request seeks for a back-up in other SN agents with similar capabilities. The selection of the best sharing alternative can be done by ranking sharing proposals (Yoon and Nof 2010; Seok and Nof 2014) or by dynamic best-matching to optimize use of available capacity (Moghaddam and Nof 2014). Results from numerical experiments show that DCS protocols can increase demand fulfillment and reduce unutilized capacity, suggesting that DCS-based flow control protocols may increase local resilience. Nevertheless, further work is required to test the performance of DCS protocols in the face of disruptions to increase resilience in SNs.

Lateral transshipments, or inventory sharing/pooling, constitute a special type of back-up sourcing in which an agent i receiving flow from a primary source may resort to inventory currently owned by an agent other than its primary sources via a sharing agreement. Paterson et al. (2011) provide a thorough review of lateral transshipment models found in literature ranging from periodic inventory redistribution models to increase protection at selected agents, to reactive models to meet demand in the case of local stock-outs. The authors' comprehensive review shows that, as in the case of DCS protocols, lateral transshipments have the potential to provide protection against disruptions. However, the available models have never been tested under a disruption environment to assess their performance in terms of local and global SN resilience.

5.1.3 Emergency Sourcing

Option contracts, presented as part of the multi-sourcing strategy in physical SNs, can also be used for emergency sourcing (Tomlin 2006; Tang and Tomlin 2008). In emergency sourcing, a SN agent i relies on a single reliable predecessor j with sufficient surplus capacity to ensure increased flow and/or expedited delivery, if needed. In order to have the possibility to adjust and/or expedite deliveries, the requesting agent must pay a premium in excess of the flow cost and may be subject to other constraints defined in the sourcing contract with agent j.

Minner (2003) reviews inventory models that consider alternative delivery modes to provide for expedited emergency deliveries; the requesting agent i can select among delivery modes with various leadtimes to balance the trade-off between increased delivery cost and the cost of disruptions due to stock-outs. From a resilience perspective, the major downside of emergency sourcing models is that these rely on a single supplier. This setting normally translates into higher sourcing costs in normal operation as well as increased exposure to unforeseen catastrophic disruptions, e.g., natural disasters, which may affect the sourcing agent.

5.1.4   The Role of Communication in Sourcing Protocols

Errors and conflicts in SN agent i's predecessors may inevitably lead to disruptions that affect input flow QoS. However, from the sourcing agent's perspective, it is important to differentiate between QoS variability and uncertainty. Variability is inherent to predecessors' processes and is beyond agent i's control. Conversely, uncertainty can be reduced, and even eliminated, by efficient and timely communication among agent i and agents $j \in P_i$.

Reducing uncertainty enables sourcing protocols to make better decisions. Gaukler et al. (2008) combine on-line order monitoring with emergency sourcing and back-up sourcing, and show through numerical tests that real-time information from predecessors regarding order status can reduce the cost of protection against disruptions. These results highlight the importance of information in reducing sourcing uncertainty and indicate the need for efficient communication protocols supported by adequately designed communication networks in order to enable cost-effective resilience in SNs.

5.2   Internal Control Protocols: Managing Resources at Agent Level

Each SN agent $a \in A$ has a set of internal resources R_a which need to be controlled and coordinated to enable transformation of input flow from predecessors into output flow sent to successors. This transformation must ensure that output flow can be delivered to each successor within the pre-defined SLA conditions. In physical SNs, internal resources normally comprise production equipment, raw material storage, WIP buffers, and finished product inventories. Several researchers have addressed the problem of coordinating production resources in order to increase throughput while minimizing WIP levels. Traditional approaches to production line flow control e.g., base-stock, kanban (Sugimori et al. 1977), and CONWIP (Spearman et al. 1990), synchronize job releases based on intermediate storage levels. Various studies analyze the performance of these methods and combinations thereof; main conclusions show that kanban outperforms base-stock by limiting the amount of WIP for a given service level (Duri et al. 2000), CONWIP optimizes the use of limited storage to achieve higher throughput than a kanban system (Pettersen and Segerstedt 2009), and a combination of kanban and CONWIP (Bonvik et al. 1997) can reduce required storage space by 20% versus kanban while achieving the same service level.

Flow control protocols for resilience operation require not only optimal performance in normal conditions but also the capacity to anticipate and avoid/overcome disruptions. When analyzed from this perspective, the aforementioned methods lack the ability to enable resilient response. Kanban was found to amplify variability emerging from machine disruptions (Bonvik et al. 1997) and to require high process reliability to achieve high throughput levels with low WIP (Deleersnyder et al. 1989). Base-stock and CONWIP lack dynamic adjustment to changing production line conditions, a shortfall partially overcome by DWIP (Yang et al. 2006). DWIP, or dynamic WIP, adjusts the CONWIP target inventory level by considering real-time leadtime of production machines, mainly affected by their operational status. Through this update protocol, DWIP is capable of overcoming disruptions and outperforms CONWIP on systems with low and high levels of disruption (Yang et al. 2006). Similarly, Ma and Koren (2004) introduce a protocol for optimal controller selection in which real-time data is used in a mathematical optimization model to select the optimal set of control rules for a production system.

Digital SN agents can, like their physical SN counterparts, comprise a variety of internal resources e.g., storage disks, routers, servers, processors, cooling systems, that need to be efficiently managed to ensure QoS delivery within pre-arranged service levels, even in the face of disruptions. For instance, data centers are increasingly required to provide on-demand service with differentiated QoS to several successors, while adaptively coping with congestion from demand surges, power shortages, equipment breakdown, and dynamic cooling requirements. Furthermore, from an economic and sustainability perspective, data centers need to provide the aforementioned service with minimum energy consumption.

Recently, research has focused on developing adaptive capacity management techniques to control the number of active host machines and the capacity allocation thereof, the size of computing clusters, and the operating frequency of each processor in a cluster. Chase et al. (2001) introduce a centralized control mechanism based on auctions and utility functions by which successors bid for service and the controller defines the optimal capacity allocation to maximize the data center's profits, i.e., utility from service provided minus cost of power to operate the data center over the allocation period. The controller actively monitors the resources' state and can dynamically re-allocate capacity under server failure and/or power shortage. Numerical experiments show that the proposed control mechanism can achieve a 25% reduction in energy consumption vs. traditional allocation methods while maintaining desired QoS.

Kusic and Kandasamy (2007) and Kusic et al. (2008) develop a control protocol for data centers under on-demand requests from successors that incorporates the possibility of adjusting processor frequency, a control variable analogous to adjusting the production rate in a production line. Using a model predictive control approach, the protocol defines the number of servers to activate, the allocation of tasks to server clusters, and the operating frequency of processors, in order to minimize power consumption while ensuring adequate QoS. Look-ahead capabilities allow the protocol to anticipate disruptions possibly emerging in a near future, enabling proactive measures to eliminate their impact. Numerical experiments show that the proposed control method can achieve around 20% reduction in energy consumption versuss a traditional control protocol while yielding the same QoS.

5.3 Distribution Protocols: Controlling Flow Delivery to Successors

With the exception of output agents $a \in A^O$, all other SN agents are required to deliver flow to one or several successors. As in the case of sourcing and internal resource operation, flow distribution may be subject to disruptions that affect QoS; examples include congestion, broken links, and interference. Therefore, resilient flow delivery requires specifically designed control protocols capable of leveraging SN structure in order to achieve connectivity and timely delivery to comply with pre-defined SLAs.

In digital SNs, signals and digital information need to be transmitted between agents i and j which may be connected by a single link (single-hop) or, most likely, through a path $\delta_{i \to j}$ formed by a series of links and nodes (multi-hop). Recently, multi-path routing techniques have received increased attention from researchers as a potential solution to deal with congestion and failed links in digital SNs. Multi-path routing consists in breaking flow to be delivered in a series of packages which are sent from agent i to agent j through a set of, ideally, fully disjoint p. paths $\Delta_{i \to j} = \left\{ \delta_{i \to j}^1, \ldots, \delta_{i \to j}^p \right\}$. The extent to which $\Delta_{i \to j}$ contains fully disjoint paths depends on the SN's level of global redundancy. It has been shown that, even if some packages are lost, the signal sent by agent i. can be reconstructed at agent j; therefore, multi-path routing can provide an energy efficient mechanism for resilient delivery of digital information (Al-Karaki and Kamal 2004).

Rohrer et al. (2009) introduce a methodology to select k-disjoint paths for $\Delta_{i \to j}$ constrained to a maximum path length, based on a path diversity metric. In this way, global redundancy can be maximally exploited while simultaneously maintaining efficiency in delivery time/energy. Numerical studies show that multi-path routing with path diversity-based seleion of $\Delta_{i \to j}$ outperforms single-path routing protocols under agent and link disruptions. Sanguesa et al. (2014) study information dissemination when SN agents are moving (constrained by road topology) and their geographical density changes dynamically. Low density leads to connectivity loss, requiring storage capabilities to retain information until it can be optimally distributed to a successor. Conversely, high density scenarios require restricted information delivery from only those agents in the most (geographically) advantageous position. The authors introduce two broadcasting protocols that can dynamically adjust to varying density conditions, avoiding connectivity loss and congestion, while increasing message delivery rations vs. traditional broadcasting protocols.

In communication networks with priority-based bandwidth allocation, Lau et al. (2008) leverage the existence of multiple paths between sources and sinks to reduce service disruptions and throughput loss due to pre-emption rules. The authors propose to re-route lower priority flow in congested links and assign bandwidth to higher priority connections, thus minimizing the number of connections dropped to free resources. Analysis of simulation results show that the proposed approach outperforms traditional pre-emption algorithms, allowing higher throughput with less disruptions.

Distribution in physical SNs also comprises single-hop, i.e. direct, delivery or multi-hop, i.e., via hub/s, delivery. As in the case of digital SNs, links (roads) and agents (hubs) in a path are subject to congestion and disruptions. Several two-stage protocols have been proposed for different types of delivery, leveraging from redundancy and real-time information. Wang and Shi (2009) and Conrad and Figliozzi (2010) study urban delivery operations e.g., package couriers and postal service, under congestion and disruptions and propose protocols that combine GIS information and real-time road network data to define an initial routing plan and dynamically re-formulate when disruption or congestion is detected, in order to minimize cost while meeting required QoS.

Fay (2000) studies the problem of train routing under dynamic customer requirements and network conditions. The author proposes a decision support system based on fuzzy logic to assist the human controller to decide what the train/s should do given a disruption/change in schedule based on real-time information. Path rating using fuzzy logic has also been proposed by Pang and Chu (2007) and Venkatasubramanian et al. (2009) to select best alternatives combining travel distance, travel time, degree of congestion, travel cost/toll and degree of difficulty.

5.4   The Need for Situation Awareness

Situation awareness (SA) has been extensively researched in the field of human factors; Endsley (1988) first defined it as the perception of the environment within a range of space and time, followed by the understanding of the elements perceived and a prognostic of their likely evolution in the near future. From this definition of situation awareness, three main components can be extracted: perception, detection, and prognostics.

Perception and detection are oftentimes addressed together. Jackson and Ferris (2013) propose that engineered systems, e.g., supply networks, should be designed following the drift correction principle which involves scanning the system to detect threats that may lead to a disruption. In physical SNs, Pettit et al. (2010)

propose to build-in a capability defined as visibility, which involves knowing the status of resources and the environment, and to provide detection methods to identify potential negative situations. Christopher and Peck (2004) point out that visibility can be increased by information exchange among a community of agents, thus highlighting the importance of an efficient communication network.

In digital SNs, Sterbenz et al. (2011) emphasize the need for context awareness, i.e., monitoring link status, e.g., congestion, and state, and the detection of adverse conditions/events in the environment. More specifically, the authors discuss the need for error detection, i.e., to identify deviations in QoS vs. expected normal operation values, and for failure detection. Smith et al. (2011) propose to: (1) design agent-level mechanisms to monitor local variables and neighboring environment and (2) detect deviations in expected QoS. Similarly, Antunes (2011) emphasizes the need for perception and awareness, system evolution monitoring, and diagnostic capabilities for exception handling to enable organizational resilience.

Although in several of the above discussed articles the words error, failure, threat, exception, adverse event, and disruption, among others, are used interchangeably, it is worth pointing out that this is a misuse of language. Moreover, in the context of situation awareness it is only appropriate to use the words fault, error, conflict, and disruption, with the following meaning:

1. *Fault* an unintended design flaw of a system or a deliberately inflicted weakness that may lead to an error (Sterbenz et al. 2010).
2. *Error* an observed or predicted deviation in any input, output, or state variable versus its expected level (Chen and Nof 2009; Sterbenz et al. 2010).
3. *Conflict* an inconsistency between SN agents' objectives, plans, or other system activities (Chen and Nof 2009).
4. *Disruption* a decline in QoS beyond minimum levels specified in a SLA, that follows an error or conflict.

In line with the previous definitions and the components of situation awareness, Chen and Nof (2009) identify five functions that can be used to build automated situation awareness into SN agents: detection, identification, isolation, diagnostics, and prognostics. These five functions call for the identification of key variables to monitor in order to detect the occurrence of an error/conflict and the design of four procedures comprising the detection of occurrence of errors/conflicts, retrieval of the location and characteristics of the error/conflict, and prognostics of possible error/conflict propagation leading to disruption.

5.5 Summary and Outlook

Researchers in physical, digital, and service networks have been increasingly addressing the need for control protocols to deal with inherent system variability. Despite the vast ground already covered, significant efforts are required to continue expanding and refining resilience-enabling protocols and to validate existing approaches from a network-level perspective.

The need for situation awareness as a resilience enabler is being increasingly recognized. Literature from digital and physical SNs shows that the integrated use of information, analytics, and decision models to forecast system evolution and define optimal allocation of excess resources, and use of redundancy are of main relevance to enabling resilience at agent level. Furthermore, as individual agent's ability to avoid disruptions improves, and recovery times are shortened, global resilience gains will be obtained from lower disruption dissemination thereby leading to more resilient SNs. Further work is required to develop generalized methods for online monitoring and selection of optimal control rules for internal resources.

Despite the contradictory results of multi-sourcing in physical SNs, outcomes from digital SNs suggest that the aforementioned strategy needs to receive more attention from researchers in the future. Analysis must be re-focused towards a resilience perspective and long-run expected behavior models, reformulated into protocols capable of adapting to, and leveraging from, dynamic conditions and information exchange among agents, as in FTTP and FTSIA. Silbermayr and Minner (2014) take an initial step towards the latter by formulating a mathematical model to derive an optimal ordering protocol under different disruption settings in a multi-sourcing environment; however, numerical analyses provided are based on cost minimization and, therefore, still lacking exploration from a resilience perspective. Future work directions must also address the combination of sourcing strategies with other flow control protocols so as to overcome the limitations and inherent inefficiencies of both strategies, thus making the combined approach more attractive than its separate components.

Examples reviewed for resilient distribution of flow, both in digital and physical SNs, highlight the importance of combining real-time data, topology information, and global redundancy to overcome connectivity problems arising from disruptions and congestion emerging dynamically. The use of redundant, parallel paths and/or the dynamic selection and re-evaluation of routing plans appear as two key components of resilient routing protocols. Further work is required to enable anticipatory capabilities that allow avoiding the occurrence of congestion by collaborative decision-making among distributed SN agents, as well as selection criteria of most adequate protocols for a given network structure.

References

Al-Karaki, J. N., & Kamal, A. E. (2004). Routing techniques in wireless sensor networks: A survey. *IEEE Wireless Communications, 11*(6), 6–28. doi:10.1109/MWC.2004.1368893

Antunes, P. (2011). BPM and exception handling: Focus on organizational resilience. *IEEE Transactions on Systems, Man, and Cybernetics, Part C (Applications and Reviews)* 41 (3): 383–392. doi:10.1109/TSMCC.2010.2062504

Asian, S., & Nie, X. (2014). Coordination in supply chains with uncertain demand and disruption risks: Existence, analysis, and insights. *IEEE Transactions on Systems, Man, and Cybernetics: Systems, 44*(9), 1139–1154. doi:10.1109/TSMC.2014.2313121

Bonvik, A. M., Couch, C. E., & Gershwin, S. B. (1997). A comparison of production-line control mechanisms. *International Journal of Production Research, 35*(3), 789–804. doi:10.1080/002075497195713

Chase, J. S., Anderson, D. C., Thakar, P. N., Vahdat, A. M., & Doyle, R. P. (2001). Managing energy and server resources in hosting centers. *ACM SIGOPS Operating Systems Review, 35* (5), 103. doi:10.1145/502059.502045

Chen, X. W., & Nof, S. Y. (2009). Automating errors and conflicts prognostics and prevention. In: Nof, S.Y. (Ed.), *Springer Handbook of Automation.* Springer Berlin Heidelberg, Berlin/Heidelberg, Germany, pp. 503–525. doi:10.1007/978-3-540-78831-7_30

Christopher, M., & Peck, H. (2004). Building the resilient supply chain. *International Journal of Logistics Management, 15*(2), 1–14. doi:10.1108/09574090410700275

Conrad, R. G., & Figliozzi, M. A. (2010). Algorithms to quantify impact of congestion on time-dependent real-world urban freight distribution networks. *Transportation Research Record: Journal of the Transportation Research Board, 2168*(1), 104–113. doi:10.3141/2168-13

Deleersnyder, J.-L., Hodgson, T. J., Muller, H., & O'Grady, P. J. (1989). Kanban controlled pull systems: An analytic approach. *Management Science, 35*(9), 1079–1091.

Duri, C., Frein, Y., & Di Mascolo, M. (2000). Comparison among three pull control policies: Kanban, base stock, and generalized kanban. *Annals of Operations Research, 93*(1–4), 41–69. doi:10.1023/A:1018919806139

Endsley, M. R. (1988). Design and evaluation for situation awareness enhancement. *Proceedings of the Human Factors and Ergonomics Society Annual Meeting, 32*(2), 97–101. doi:10.1177/154193128803200221

Fay, A. (2000). A fuzzy knowledge-based system for railway traffic control. *Engineering Applications of Artificial Intelligence, 13*(6), 719–729. doi:10.1016/S0952-1976(00)00027-0

Federgruen, A., & Yang, N. (2008). Selecting a portfolio of suppliers under demand and supply risks. *Operations Research, 56*(4), 916–936. doi:10.1287/opre.1080.0551

Fu, Q., Lee, C.-Y., & Teo, C.-P. (2010). Procurement management using option contracts: Random spot price and the portfolio effect. *IIE Transactions, 42*(11), 793–811. doi:10.1080/07408171003670983

Gaukler, G. M., Özer, Ö., & Hausman, W. H. (2008). Order progress information: Improved dynamic emergency ordering policies. *Production and Operations Management, 17*(6), 599–613. doi:10.3401/poms.1080.0066

Glock, C. H., & Ries, J. M. (2013). Reducing lead time risk through multiple sourcing: The case of stochastic demand and variable lead time. *International Journal of Production Research, 51* (1), 43–56. doi:10.1080/00207543.2011.644817

Hill, R. M. (1996). Order splitting in continuous review (Q, r) inventory models. *European Journal of Operational Research, 95*(1), 53–61. doi:10.1016/0377-2217(95)00257-X

Jackson, S., & Ferris, T. L. J. (2013). Resilience principles for engineered systems. *Systems Engineering, 16*(2), 152–164. doi:10.1002/sys.21228

Kusic, D., & Kandasamy, N. (2007). Risk-aware limited lookahead control for dynamic resource provisioning in enterprise computing systems. *Cluster Computing, 10*(4), 395–408. doi:10.1007/s10586-007-0022-y

Kusic, D., Kephart, J. O., Hanson, J. E., Kandasamy, N., & Jiang, G. (2008). Power and performance management of virtualized computing environments via lookahead control. *Cluster Computing, 12*(1), 1–15. doi:10.1007/s10586-008-0070-y

Lau, C. H., Soong, B.-H., & Bose, S. K. (2008). Preemption with rerouting to minimize service disruption in connection-oriented networks. *IEEE Transactions on Systems, Man, and Cybernetics - Part A: Systems and Humans, 38*(5), 1093–1104. doi:10.1109/TSMCA.2008.2001075

Liu, Y., & Nof, S. Y. (2004). Distributed microflow sensor arrays and networks: Design of architectures and communication protocols. *International Journal of Production Research, 42* (15), 3101–3115. doi:10.1080/00207540410001699363

Liu, Y., & Nof, S. Y. (2008). Fault-tolerant sensor integration for micro flow-sensor arrays and networks. *Computers & Industrial Engineering, 54*(3), 634–647. doi:10.1016/j.cie.2007. 09.013

Ma, Y.-H., & Koren, Y. (2004). Operation of manufacturing systems with work-in-process inventory and production control. *CIRP Annals—Manufacturing Technology, 53*(1), 61–365. doi:10.1016/S0007-8506(07)60717-3

Minner, S. (2003). Multiple-supplier inventory models in supply chain management: A review. *International Journal of Production Economics, 81–82,* 265–279. doi:10.1016/S0925-5273 (02)00288-8

Moghaddam, M., & Nof, S. Y. (2014). Combined demand and capacity sharing with best matching decisions in enterprise collaboration. *International Journal of Production Economics, 148,* 93–109. doi:10.1016/j.ijpe.2013.11.015

Nof, S. Y. (2003). Design of effective e-Work: Review of models, tools, and emerging challenges. *Production Planning & Control, 14*(8), 681–703. doi:10.1080/09537280310001647832

Nof, S. Y. (2007). Collaborative control theory for e-Work, e-Production, and e-Service. *Annual Reviews in Control, 31*(2), 281–292. doi:10.1016/j.arcontrol.2007.08.002

Pan, A. C., Ramasesh, R. V., Hayya, J. C., & Keith Ord, J. (1991). Multiple sourcing: The determination of lead times. *Operations Research Letters, 10*(1), 1–7. doi:10.1016/0167-6377 (91)90079-5

Pang, G., & Chu, M.-H. (2007). Route selection for vehicle navigation and control. In: *Proceedings of the 33rd Annual Conference of the IEEE Industrial Electronics Society.* Hong Kong, China, pp. 693–698. doi:10.1109/IECON.2007.4459980

Paterson, C., Kiesmüller, G., Teunter, R., & Glazebrook, K. (2011). Inventory models with lateral transshipments: A review. *European Journal of Operational Research, 210*(2), 125–136. doi:10.1016/j.ejor.2010.05.048

Pettersen, J.-A., & Segerstedt, A. (2009). Restricted work-in-process: A study of differences between Kanban and CONWIP. *International Journal of Production Economics, 118*(1), 199–207. doi:10.1016/j.ijpe.2008.08.043

Pettit, T. J., Fiksel, J., & Croxton, K. L. (2010). Ensuring supply chain resilience: Development of a conceptual framework. *Journal of Business Logistics, 31*(1), 1–21. doi:10.1002/j.2158-1592. 2010.tb00125.x

Rohrer, J. P., Jabbar, A., & Sterbenz, J. P. G. (2009). Path diversification: A multipath resilience mechanism. In: *Proceedings of the 7th International Workshop on Design of Reliable Communication Networks.* Washington, DC, USA, pp. 343–351. doi:10.1109/DRCN.2009. 5339988

Sanguesa, J. A., Fogue, M., Garrido, P., Martinez, F. J., Cano, J., & Calafate, C. T. (2014). Using topology and neighbor information to overcome adverse vehicle density conditions. *Transportation Research Part C: Emerging Technologies, 42,* 1–13. doi:10.1016/j.trc.2014. 02.010

Seok, H., & Nof, S. Y. (2014). Collaborative capacity sharing among manufacturers on the same supply network horizontal layer for sustainable and balanced returns. *International Journal of Production Research, 52*(6), 1622–1643. doi:10.1080/00207543.2013.842016

Silbermayr, L., & Minner, S. (2014). A multiple sourcing inventory model under disruption risk. *International Journal of Production Economics, 149,* 37–46. doi:10.1016/j.ijpe.2013.03.025

Smith, P., Fessi, A., Lac, C., Hutchison, D., Sterbenz, J. P. G., Scholler, M., et al. (2011). Network resilience: A systematic approach. *IEEE Communications Magazine, 49*(7), 88–97. doi:10. 1109/MCOM.2011.5936160

Spearman, M. L., Woodruff, D. L., & Hopp, W. J. (1990). CONWIP: A pull alternative to kanban. *International Journal of Production Research, 28*(5), 879–894.

Sterbenz, J. P. G., Cetinkaya, E. K., Hameed, M. A., Jabbar, A., Qian, S., & Rohrer, J. P. (2011). Evaluation of network resilience, survivability, and disruption tolerance: Analysis, topology generation, simulation, and experimentation. *Telecommunication Systems, 52*(2), 705–736. doi:10.1007/s11235-011-9573-6

Sterbenz, J. P. G., Hutchison, D., Çetinkaya, E. K., Jabbar, A., Rohrer, J. P., Schöller, M., et al. (2010). Resilience and survivability in communication networks: Strategies, principles, and survey of disciplines. *Computer Networks, 54*(8), 1245–1265.

Sugimori, Y., Kusunoki, K., Cho, F., & Uchikawa, S. (1977). Toyota production system and kanban system. Materialization of just-in-time and respect-for-human system. *International Journal of Production Research, 15*(6), 553–564. doi:10.1080/00207547708943149

Tang, C. S., & Tomlin, B. (2008). The power of flexibility for mitigating supply chain risks. *International Journal of Production Economics, 116*(1), 12–27. doi:10.1016/j.ijpe.2008.07.008

Thomas, D. J., & Tyworth, J. E. (2006). Pooling lead-time risk by order splitting: A critical review. *Transportation Research Part E: Logistics and Transportation Review, 42*(4), 245–257. doi:10.1016/j.tre.2004.11.002

Tomlin, B. (2006). On the value of mitigation and contingency strategies for managing supply chain disruption risks. *Management Science, 52*(5), 639–657. doi:10.1287/mnsc.1060.0515

Venkatasubramanian, S. N., Vaidyanathan, G., & Duraisamy, S. (2009) Fuzzy logic based navigation system. In: *Proceedings of the Third International Symposium on Intelligent Information Technology Application.* Nanchang, China, pp. 69–72. doi:10.1109/IITA.2009.340

Wang, Z., & Shi, J. (2009). A model for urban distribution system under disruptions of vehicle travel time delay. In: *Proceedings of the 2nd International Conference on Intelligent Computing Technology and Automation,* Dalian, China, pp. 433–436. doi:10.1109/ICICTA.2009.570

Yang, R. L., Subramaniam, V., & Gershwin, S. B. (2006). *Setting real time WIP levels in production lines.* Innovation in Manufacturing Systems and Technology (IMST): Singapore-MIT Alliance.

Yoon, S. W., & Nof, S. Y. (2010). Demand and capacity sharing decisions and protocols in a collaborative network of enterprises. *Decision Support Systems, 49*(4), 442–450. doi:10.1016/j.dss.2010.05.005

Chapter 6
Resilience by Teaming Framework

Over the last decade, researchers and practitioners have been increasingly focusing on developing strategies to create resilient supply networks. Most approaches focus on optimizing the trade-off between operational costs and resilience through increased protection, in the form of redundancy and excess resources, and/or the use of more reliable agents/resources. Nature and humans, however, show us that it is also possible to create resilient systems with fault-prone agent and resources through smart designs and distributed control protocols. The driving force behind these resilient systems is *teaming*.

Teaming is a process by which a set of agents form a network and collaborate to achieve their individual goals and, perhaps, a common objective. Inspired in the Fault-tolerance by Teaming principle of Collaborative Control Theory (Nof 2007), the Resilience by Teaming framework (Reyes Levalle 2015; Reyes Levalle and Nof 2015) comprises a set of principles and protocols to design and operate resilient supply network on the basis of forming collaborative teams among fault-prone agents. Leveraging the mathematical formalism to describe and model supply networks from Chap. 2, the set of fundamentals for resilience in supply networks from Chap. 3, and the basic strategies to design and operate resilient SNs from Chaps. 4 and 5, the following sections introduce the main components of the Resilience by Teaming framework. Chapters 7 through 9 discuss each component of the Resilience by Teaming framework in detail.

6.1 The Need for Resilience Through Collaboration and Teaming

SN agents' interactions constitute a form of e-work, defined by Nof (2003) as *any collaborative, computer-supported and communication-enabled productive activities in highly distributed organizations of humans and/or robots or autonomous*

© Springer International Publishing AG 2018
R. Reyes Levalle, *Resilience by Teaming in Supply Chains and Networks*,
Automation, Collaboration, & E-Services 5, DOI 10.1007/978-3-319-58323-5_6

Table 6.1 FTT fundamentals and their link to resilience dimensions

FTT fundamental	Definition	Aspects	Resilience dimension
Team formation	Dynamic selection of team participants from a finite population of agents (e.g., Jeong and Nof 2009)	• Centralized (one agent selects team members) or de-centralized (two or more agents by consensus or group decision-making) • Team size and member responsibilities; topology, redundancy • Dynamic association/dissociation	SN structure
Collaboration among team members	Real-time control of communication and flow exchanges, and resource sharing among team members (e.g., Liu and Nof (2004, 2008) and Yoon and Nof (2010))	• Centralized (team leader) versus de-centralized (consensus) decisions • Collaboration requirements and constraints • Collaboration type: lateral vs. vertical, information/flow exchange, resource sharing	SN control protocols

systems. Over the last decades, several researchers have collaborated to develop and refine a set of six principles to design e-work systems, leading to the emergence of Collaborative Control Theory (Nof 2007). Although each of the CCT principles can have a meaningful impact on the design and control of resilient SNs, the principle of Fault-Tolerance by Teaming stands out as a potential enabler of resilient performance among agents susceptible to disruptions.

FTT principle is based on the notion that a team of weaker agents can outperform single flawless agent by enabling smart automation to overcome (temporarily) faulty agents (Nof 2007; Velásquez and Nof 2009). Although necessary for resilience, fault-tolerance is not a sufficient condition for resilient SNs (Sterbenz et al. 2010). Nevertheless, the fundamentals of FTT (Table 6.1), combined with notions from the Conflict and Error Detection and Prognostics principle of CCT, can be extended and generalized into a Resilience by Teaming framework, to enable the formation, re-configuration, and operation of SNs of disruption-prone agents capable of achieving higher resilience than a SN composed by flawless (or more reliable) agents.

6.2   An Overview of the Resilience by Teaming Framework

Supply network topology and resilience emerge from local, i.e., agent-level, decisions. Although these decisions may be influenced by network-level interactions and collaboration with other agents, each SN agent is responsible for designing and controlling its own internal resource network, as well as defining sourcing and distribution networks to enable inter-agent flow. The Resilience by Teaming framework (Fig. 6.1) defines two network-level activities, situation awareness and agent-agent negotiation, and two agent-level activities, team formation (design) and collaboration for resilience (operation), in order to enable SN agents to be locally resilient while contributing to the creation of a globally resilient supply networks.

Situation awareness requires each SN agent to participate in privacy-preserving information exchanges with other agents and SN regulators to collaboratively detect vulnerabilities in topology and/or agent interactions. SN regulators are a special type of agent which determines environment constraints to individual agent behavior, for all agents in a SN, to ensure fairness and help protect the SN against disruptions. For instance, government agencies defining antitrust regulations are SN regulators.

The collective level of situation awareness depends on the efficiency of the communication network established by SN agents to disseminate relevant information for decision making. Different levels of situation awareness are possible, based on agent-level situation awareness and the characteristics of the communication network; nevertheless, RBT protocols do not explicitly place minimum requirements on situation awareness. Furthermore, while situation awareness can be measured (Salmon et al. 2006; Wildman et al. 2014), RBT protocols do not quantify SA. Each agent must define with whom to set communication links, beyond those agents with which it engages in flow exchange. Information obtained directly from other agents or derived from collaborative analysis of data will affect decisions made locally with respect to predecessor selection, as shown in Fig. 6.1.

The second network-level activity, agent-agent negotiation, comprises all strategic decisions made collaboratively among agents to define topological requirements, share excess flow demand and internal capacity, and define SLAs, with the objective of enhancing their individual and collective performance and resilience.

Agent-level team design decisions are influenced by network-level context and information. RBT framework comprises two agent-level team design protocols, sourcing team formation/re-configuration protocol (STF/RP) and distribution network formation/re-configuration protocol (DNF/RP), to dynamically define sourcing and distribution networks, respectively, and two agent-level design guidelines for internal resource networks to enable resilience via resource teaming. Following FTT fundamentals, STF/RP selects a team of weak sourcing agents to provide primary flow supply and establishes collaboration with a secondary set of weak agents to create flow sharing conditions in order to overcome lack of flow from primary sources. DNF/RP dynamically evaluates delivery network conditions to define the best performing team to execute delivery, based on disruption protection and cost. Design of internal resource networks is application-specific; nonetheless, guidelines on process parallelism and storage design for active protective capacity management enable resource teaming to avoid/overcome disruptions at a lower operational and, possibly, design cost.

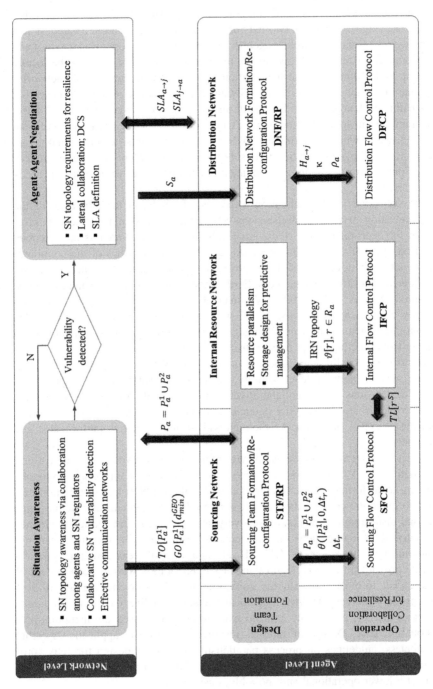

Fig. 6.1 Resilience by Teaming framework

Team formation decisions and collaborative operation for resilience in sourcing, internal flow, and distribution are interrelated. Selected topologies and team participant characteristics may enhance, or hinder, operational protocols' ability to provide resilient performance. SN agents apply the sourcing flow control protocol (SFCP) to regulate ordering from primary and secondary sourcing teams, leveraging teaming capabilities within P_a. Selection of replenishment parameter Δt_r, in turn, affects team selection. The internal flow control protocol (IFCP) is used by SN agents to manage internal resources R_a in order to ensure steady input-to-output transformation, minimizing variability arising from disruptions. The protocol is highly dependent on the internal resource network's capacity to enable teaming of processes, and processes and storages. Furthermore, dynamic selection of internal storage parameters may affect SFCP, as replenishment orders may be placed with higher or lower frequency. Finally, the distribution flow control protocol (DFCP) manages flow delivery from agent a to its successors S_a. As in the case of SFCP, decisions made during team formation constrain DFCP's ability of provide resilient response to potential, or actual, disruptions. Furthermore, re-teaming threshold P_a may affect decisions made during team formation, as lower thresholds will require more frequent re-configurations.

The above discussed protocols of the Resilience by Teaming framework are discussed in detail in the subsequent chapters.

References

Jeong, W., & Nof, S. Y. (2009). A collaborative sensor network middleware for automated production systems. *Computers & Industrial Engineering, 57*(1), 106–113. doi:10.1016/j.cie.2008.11.007.

Liu, Y., & Nof, S. Y. (2004). Distributed microflow sensor arrays and networks: Design of architectures and communication protocols. *International Journal of Production Research, 42*(15), 3101–3115. doi:10.1080/00207540410001699363

Liu, Y., & Nof, S.Y., (2008). Fault-tolerant sensor integration for micro flow-sensor arrays and networks. *Computers & Industrial Engineering, 54*(3), 634–647. doi:10.1016/j.cie.2007.09.013

Nof, S. Y. (2003). Design of effective e-work: Review of models, tools, and emerging challenges. *Production Planning & Control, 14*(8), 681–703. doi:10.1080/09537280310001647832

Nof, S. Y. (2007). Collaborative control theory for e-work, e-production, and e-service. *Annual Reviews in Control, 31*(2), 281–292. doi:10.1016/j.arcontrol.2007.08.002

Reyes Levalle, R. (2015). *Resilience by teaming in supply networks.* West Lafayette: Purdue University.

Reyes Levalle, R., & Nof, S. Y. (2015). A resilience by teaming framework for collaborative supply networks. *Computers & Industrial Engineering, 90,* 67–85. doi:10.1016/j.cie.2015.08.017

Salmon, P., Stanton, N., Walker, G., & Green, D. (2006). Situation awareness measurement: A review of applicability for C4i environments. *Applied Ergonomics, 37,* 225–238. doi:10.1016/j.apergo.2005.02.001

Sterbenz, J. P. G., Hutchison, D., Çetinkaya, E. K., Jabbar, A., Rohrer, J. P., Schöller, M., et al. (2010). Resilience and survivability in communication networks: Strategies, principles, and survey of disciplines. *Computer Networks, 54*(8), 1245–1265. doi:10.1016/j.comnet.2010.03. 005

Velásquez, J.D., and Nof, S.Y. 2009. Collaborative e-Work, e-Business, and e-Service, In Nof, S. Y. (Ed.), *Springer handbook of automation* (pp. 1549–1576). Springer, Berlin, doi:10.1007/ 978-3-540-78831-7_88

Wildman, J. L., Salas, E., & Scott, C. P. R. (2014). Measuring cognition in teams: A cross-domain review. *Human Factors, 56*(5), 911–941. doi:10.1177/0018720813515907

Yoon, S. W., & Nof, S. Y. (2010). Demand and capacity sharing decisions and protocols in a collaborative network of enterprises. *Decision Support Systems, 49*(4), 442–450. doi:10.1016/j. dss.2010.05.005

Chapter 7
Resilience by Teaming: Sourcing Network Design and Flow Management Protocols

Every SN agent a in $A^K \cup A^O$ needs to receive input flow from a set of predecessors P_a in order to operate and provide output to its successors. Such agents face a medium-term strategic decision, i.e., which agents to select as P_a, and a short-term operational decision, i.e., which agents in P_a to source from at time t. These distributed strategic decisions of agents $a \in A$ shape the dynamic topology of the supply network, and give rise to its structural resilience, as well as determine, at least in part, the ability of each SN agent to cope with sourcing disruptions.

This chapter presents the theoretical aspects of two sourcing protocols inspired in the Fault-tolerance by Teaming principle, FTT, of Collaborative Control Theory (Nof 2007), CCT, to select a team of predecessors P_a and to adaptively manage flow from these agents. The Sourcing Team Formation/Re-configuration Protocol (STF/RP) and Sourcing Flow Control Protocol (SFCP) leverage inherent randomness in individual agents in P_a to provide a more stable and resilient input to agent a.

7.1 Lessons Learnt from the FTT Principle of CCT Applied to Sensor Networks

In WSNs, the selection of predecessors and transmission protocols by base stations affects sensors energy consumption and network response time. In general, network structures with higher connectivity enable transmission protocols to exploit redundancy, thus creating fault-tolerance. However, increased number of transmissions leads to higher energy consumption and, consequently, shorter sensor battery lives. This, in turn, undermines fault-tolerance and calls for frequent network re-configurations, which accelerate the rate of energy consumption.

© Springer International Publishing AG 2018
R. Reyes Levalle, *Resilience by Teaming in Supply Chains and Networks*,
Automation, Collaboration, & E-Services 5, DOI 10.1007/978-3-319-58323-5_7

Over time, several network architectures and transmission protocols have been explored in an attempt to increase energy efficiency and response times, while enabling fault-tolerance (see Wesson et al. (1981), Iyengar et al. (1994), and Qi et al. (2001) for a survey of sensor network architectures, and Al-Karaki and Kamal (2004) and Wan et al. (2008) for a survey of transmission protocols). Liu and Nof (2004) and Jeong and Nof (2009) apply the FTT principle of CCT to address predecessor selection in WSNs. Liu and Nof (2004) argue that, in order to enable fault-tolerance via FTTP, a flow control protocol that can leverage redundancy, the network formed by a base station and its predecessors must follow a cluster network architecture (CLA), shown in Fig. 7.1.

In CLA, a base station a of hierarchy 1 is directly connected to a set of predecessors P_a of hierarchy 0 from which it receives flow after a sourcing request. Similarly, base stations of hierarchy z are arranged in cluster formations to supply a base station of hierarchy $z + 1$, until reaching the highest hierarchy base station, e.g., a CPU. When a network is arranged following a CLA and sourcing is controlled by the fault-tolerance time-out protocol, FTTP, any request made to agent $i \in P_a$ that remains unfulfilled after a predefined time-out is re-assigned to a contingency agent $j \in P_a$, $i \neq j$. In order for the original (undelivered) message from i to a to reach destination through j, the latter must have received the original broadcast from i. Then, agent i use a flooding protocol or a gossiping protocol with more than two receivers to send messages to its successors $j \in S_i$.

In order to balance energy consumption and extend sensors' battery life, Jeong and Nof (2009) developed a simulated annealing algorithm to dynamically form and dissolve sensor clusters. Team formation is controlled by a central agent, i.e., the highest hierarchy base station, which defines P_a for every base station a in the network such that there are no overlaps between different P_a.

Despite the benefits achieved by CLA, FTTP, and the simulated annealing clustering algorithm, these present several limitations (Table 7.1) and are only applicable to WSNs; they cannot be directly applied to other types of supply networks.

The next sections introduce two protocols based on the FTT principle of CCT: (1) Sourcing Team Formation/Re-configuration Protocol (STF/RP) and (2) Sourcing Flow Control Protocol (SFCP). These protocols are inspired by insights from the

Fig. 7.1 Cluster network architecture, CLA

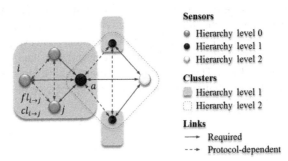

Sensors

◒ Hierarchy level 0
● Hierarchy level 1
◔ Hierarchy level 2

Clusters

▨ Hierarchy level 1
▢ Hierarchy level 2

Links

⟶ Required
--⟶ Protocol-dependent

Table 7.1 Limitations of CLA, FTTP, and FTT-based re-clustering algorithm

Limitation	Description	Needs for applicability to other SNs
Dual role of agents in P_a	• CLA assumes agents in P_a are able to source in normal operation and when requested by FTTP after a time-out	• P_a formed by two teams of agents: (i) P_a^1, responsible for sourcing in normal operation (ii) P_a^2, responsible for contingent sourcing when agent $i \in P_a^2$ is not able to source
CLA + FTTP cannot overcome faulty agents	• CLA + FTTP were designed to cope only with faulty links • Re-clustering partially addresses the problem; high re-clustering frequency affects energy efficiency	• Build-in redundancy in P_a; not all $i \in P_a$ required to source a simultaneously • Sourcing protocols capable of leveraging redundancy in P_a to cope with agent disruption without need for frequent re-configurations
A link $i \rightarrow j$ exists iff i and j belong to the same cluster; agents belong to, at most, 2 clusters	• CLA is a fully connected network lacking links to $i \notin \{a\} \cup P_a, i \neq a$ • Contingency agents must be supplied by agents in P_a	• SN agents must be able to form links beyond their participation in a given team P_a • Less structured network formation and re-configuration; higher path diversity • Cotingency flow must be equivalent to normal flow, but not necessarily come from the same agents in P_a
Restricted information sharing among agents	• Energy constraints in WSNs applications limit communication among agents	• Flow control protocols must maximize use of available information for CEDP

works of Liu and Nof (2004) and Jeong and Nof (2009) and overcome the limitations presented in Table 7.1, extending their applicability to all SNs.

7.2 Sourcing Team Formation/Re-configuration Protocol (STF/RP)

7.2.1 Formalism of Sourcing Teams

In CLA, predecessors of agent a have a dual role: (1) to source a directly, and (2) to act as a relay for flow coming from their sibling nodes, i.e., other predecessors of

a within the cluster. In general, following the observations summarized in Table 7.1, two teams of predecessors can be defined based on their interaction with agent a:

(1) Primary sourcing team (T1): A set P_a^1 of agents $i \in A$ that send flow to a on a regular basis, upon a's request, but do not receive flow from a. Formally,

$$P_a^1 = \{i | i \in P_a \wedge \nexists fl_{a \to i}\}$$

(2) Secondary sourcing team (T2): A set P_a^2 of agents $j \in A$ that send flow to a on a regular basis, upon a's request, and may require similar sourcing from a, upon request. Formally,

$$P_a^2 = \{j | j \in P_a \wedge a \in P_j\}$$

Any SN agent sourcing agent a must belong to, at least, one sourcing team, T1 or T2. Then,

$$P_a^1 \cup P_a^2 = P_a$$

and agents with dual role exist if and only if $P_a^1 \cap P_a^2 \neq \emptyset$. It follows this definition that CLA is a special case where $P_a^1 \cap P_a^2 = P_a$.

7.2.2　Design of the Primary Sourcing Team

The primary sourcing team, T1, must be designed by selecting a set of agents P_a^1 which are collectively capable to overcome individual agent disruptions and/or congestion. In part, their collective ability to achieve such performance will be determined by the protocol that controls sourcing flow (SFCP, Sect. 7.3); however, participant selection will limit the extent to which SFCP can leverage teaming to achieve sourcing resilience.

Consider a set of agents $i \in A$ capable of providing agent a with input flow and let Δt_i be the time required by agent i to deliver flow to agent a, upon request from the latter. From a design perspective, Δt_i is a random variable which follows a known probability distribution, with pdf ψ_i and cdf Ψ_i. Therefore, $\Psi_i[\Delta t_i \leq \Delta t_r]$ is the probability that agent i delivers flow to agent a within a predefined delivery time Δt_r (Fig. 7.2).

SFCP is based on a bidding mechanism by which all agents in P_a^1 may submit a sourcing proposal upon request from agent a, with delivery time distribution Ψ_i. Through SFCP, agents in P_a^1 team to provide at least one bid with $\Delta t_i \leq \Delta t_r$ in each

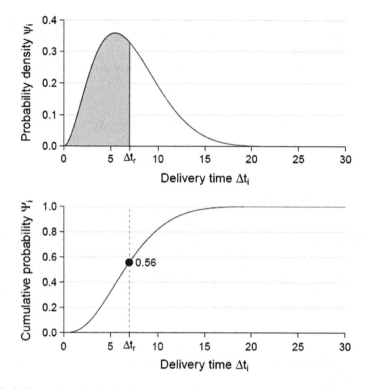

Fig. 7.2 Delivery time distribution for agent i—Example

bidding round, with high probability θ^*. Let $\theta(w, x, \Delta t_r)$ be the probability that x agents out of a team of w provide a bid with $\Delta t_i \leq \Delta t_r$, then,

$$\theta^* \leq 1 - \theta(w^*, 0, \Delta t_r)$$

for some team P_a^1 such that $w^* = |P_a^1|$.

The expression for $\theta(w^*, 0, \Delta t_r)$ depends on the assumptions made about ψ_i. If agents i delivery times are iid, the probability of having no deliveries within Δt_r from a team of w participants is

$$\theta(w, 0, \Delta t_r) = \binom{w}{0} (p(\Delta t_r))^0 (1 - p(\Delta t_r))^{w-0} = (1 - p(\Delta t_r))^w$$

where $\Psi_i[\Delta t_i \leq \Delta t_r] = p(\Delta t_r), \forall i$. Then,

$$\theta^* \leq 1 - \theta(w^*, 0, \Delta t_r) = 1 - (1 - p(\Delta t_r))^{w^*} \rightarrow w^* = \left\lceil \frac{\ln 1 - \theta^*}{\ln 1 - p(\Delta t_r)} \right\rceil$$

where $\lceil \rceil$ is the ceiling function. Hence, in order to select an optimal team P_a^1 of iid agents, it suffices to select any w^* agents i, and w^* is completely defined by the parameters θ^* and Δt_r.

In the case agents i delivery time is not identically distributed, the expression becomes

$$\theta^* \leq 1 - \theta(w^*, 0, \Delta t_r) = 1 - \prod_{i=1}^{w^*}(1 - \Psi_i[\Delta t_i \leq \Delta t_r])$$

$$\rightarrow \ln 1 - \theta^* \geq \sum_{i=1}^{w^*} \ln(1 - \Psi_i[\Delta t_i \leq \Delta t_r])$$

Then, it suffices to select any team of w^* agents that satisfies the condition above. Note that, the expression on the right side is monotonically decreasing with w^*, as $\ln(1 - \Psi_i[\Delta t_i \leq \Delta t_r])$ is always lower than or equal to 0. Hence, a lower bound w_{min}^* for w^* can be obtained by ranking agents in decreasing order of $\Psi_i[\Delta t_i \leq \Delta t_r]$ and selecting the top w^* agents. Also note that any team $w^* > w_{min}^*$ that satisfies the design condition will be composed of weaker agents than w_{min}^* but capable of achieving the same collective output.

7.2.3   The Effect of Correlation on Primary Sourcing Team Selection

Lack of independence among delivery times from agents i may arise from several factors, as outlined in Sect. 4.2.2, and can only be assessed by a data-driven analysis of joint probability distributions. Nevertheless, P_a^1 can be designed to minimize the incidence of topological and topographical aspects possibly leading to correlation among ψ_i.

Topological dependency originates in shared sourcing structures among agents i, a condition that may induce correlation in lack of input flow due to disruptions to a common predecessor. Let $TO[P_a^1]$ be the index of topological overlap between the predecessors of agents $i \in P_a^1$, obtained as follows:

$$TO[P_a^1] = \frac{\sum_{i \in P_a^1} deg^{IN}(i) - \left| \bigcup_{i \in P_a^1} P_i^1 \right|}{\left| \bigcup_{i \in P_a^1} P_i^1 \right| (|P_a^1| - 1)}$$

$TO[P_a^1] \in [0,1]$ measures the ratio between (1) the number of overlapping flow links and (2) the number of flow links in a completely overlapped network, between agents $i \in P_a^1$ and their corresponding predecessors P_i^1. A $TO[P_a^1]$ index of 0 implies no correlation emerging from overlap, as sourcing structures are completely

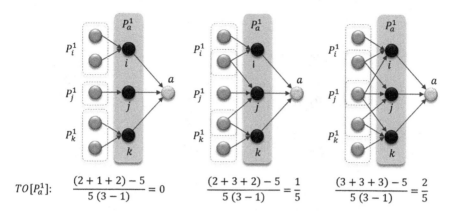

$$TO[P_a^1]: \qquad \frac{(2+1+2)-5}{5\,(3-1)}=0 \qquad\qquad \frac{(2+3+2)-5}{5\,(3-1)}=\frac{1}{5} \qquad\qquad \frac{(3+3+3)-5}{5\,(3-1)}=\frac{2}{5}$$

Fig. 7.3 Index of topological overlap—Example

independent while an index of 1 implies a full overlap in the sourcing structures, inflicting maximum correlation possibility among agents i. Figure 7.3 illustrates the relationship between $TO[P_a^1]$ and various alternative topologies.

Topographical correlation, on the other hand, depends on the relative location of agents. Proximity may render groups of agents incapable of avoid fate-sharing when the disruption originates from location-based disruptions e.g., natural disasters, political instability, interference (in WSNs), labor strikes, power outages. Let d_{min}^{GEO} be a parameter describing the minimum distance required to avoid topographical correlation between any two agents $i,j \in P_a^1$ and let $\Gamma(i, d_{min}^{GEO})$ be the number of agents within d_{crit} of agent i. Then, the index of topographical overlap $GO[P_a^1](d_{min}^{GEO})$ is defined as

$$GO[P_a^1](d_{min}^{GEO}) = \frac{\sum_{i \in P_a^1} \Gamma(i, d_{min}^{GEO})}{|P_a^1|(|P_a^1|-1)}$$

where $GO[P_a^1](d_{min}^{GEO}) \in [0,1]$. $GO[P_a^1](d_{min}^{GEO}) = 0$ if no agent has neighbors within d_{min}^{GEO}, and $GO[P_a^1](d_{min}^{GEO}) = 1$ if all agents $i \in P_a^1$ are enclosed in a circumference of radius d_{min}^{GEO}. Figure 7.4 illustrates the application of $GO[P_a^1](d_{min}^{GEO})$.

Using $TO[P_a^1]$ and $GO[P_a^1](d_{min}^{GEO})$ it is possible to rank a set of teams based on the compound index $TO[P_a^1] + \upsilon\,GO[P_a^1](d_{min}^{GEO})$, where parameter υ, selected by agent a, defines the relative importance of topological vs. topographical overlap among team participants. In order to ensure privacy among agents i, which may not be willing to provide information regarding their predecessors to agent a, a SN regulator may provide agent a with a list of teams ranked in decreasing order of $TO[P_a^1] + \upsilon\,GO[P_a^1](d_{min}^{GEO})$, so that agent a may later select the team that fulfills the non-correlated design criterion. Once team members are selected, the delivery time distribution ψ_i becomes part of the SLA between agent i and a.

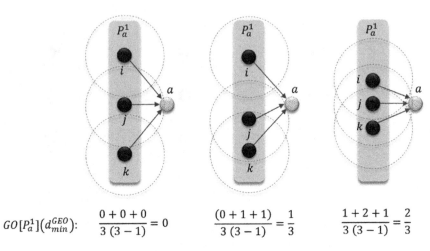

$$GO[P_a^1](d_{min}^{GEO}):\qquad \frac{0+0+0}{3\,(3-1)}=0 \qquad\qquad \frac{(0+1+1)}{3\,(3-1)}=\frac{1}{3} \qquad\qquad \frac{1+2+1}{3\,(3-1)}=\frac{2}{3}$$

Fig. 7.4 Index of topographical overlap—Example

7.2.4 Design of the Secondary Sourcing Team

The secondary team, P_a^2, is required to address only a small fraction, more precisely $1-\theta^*$, of the sourcing needs emerging from agent a. Per design of SCFP, the secondary team's responsibility is to enable flow sharing among a community of agents that can benefit from such exchange to cover sourcing requirements not addressed by P_a^1. Most of the time the structure formed by P_a^2 should remain dormant, implying little, if any, resource consumption for the agents involved. Therefore, agent a must select as many agents i as possible (given any resource constraints) to form P_a^2.

7.2.5 Sourcing Team Formation/Re-configuration Protocol (STF/RP)

Design considerations discussed in the above sections are implemented by the Sourcing Team Formation/Re-configuration Protocol (STF/RP). The interaction steps between agent a, agents i that can potentially source a, and the network regulator are presented in Fig. 7.5. Protocol decisions are adjusted by tuning six parameters described in Table 7.2.

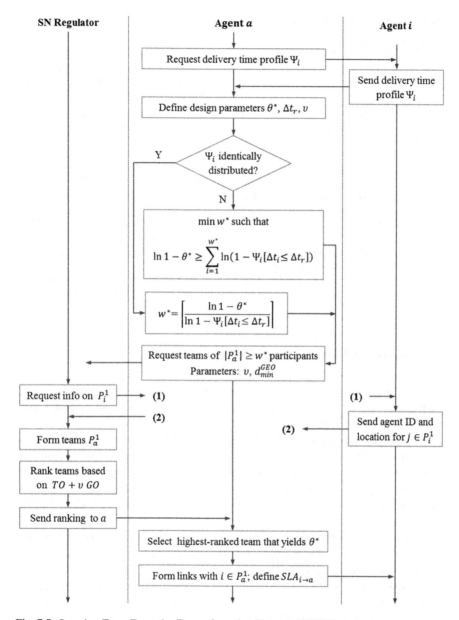

Fig. 7.5 Sourcing Team Formation/Re-configuration Protocol (STF/RP)

Table 7.2 STF/RP parameters and effects thereof

Parameter	Description	Effect on STF/RP
Δt_r	Look-ahead time for sourcing gap detection (in SFCP)	• Selection affects sourcing network size • Higher values require smaller P_a^1 (for a given service level)—but may increase the frequency of use of P_a^2
θ^*	Probability of receiving at least one bid within Δt_r from $i \in P_a^1$ (in SFCP)	• Affects the relative frequency between sourcing from primary and secondary teams • Higher values lead to larger primary teams and, possibly, more flow from agents $i \in P_a^1$ versus $j \in P_a^2$
υ	Relative importance of topological vs. topographical overlap among $i \in P_a^1$	• Affects ranking of possible sourcing teams • Higher values give more relevance to topographical overlap
d_{min}^{GEO}	Minimum distance to avoid topographical correlation between any two agents $i, j \in P_a^1$	• Controls criterion for topographical overlap • Higher values increase probability of overlap, affecting possible sourcing teams ranking

7.3   Sourcing Flow Control Protocol (SFCP)

The sourcing protocol by which agent a manages sourcing from its predecessors P_a must enable the following collaboration and teaming capabilities to foster resilience: (1) teaming among agents in P_a^1 to meet delivery time Δt_r (2) information exchange regarding pending delivery status for updated sourcing request control, increasing visibility of congestion and disruption, (3) prompt communication of disruptions from $i \in P_a^1$ to a, to initiate any required recovery actions, and (4) acceptance of overlapping deliveries, to increase protection of agent a and allow continuity of operations at agents i, even when they offer long or highly variable leadtimes.

Let a be a SN agent that can receive flow from, and communicate with, agents $i \in P_a^1$ and $j \in P_a^2$. In order to evaluate future sourcing needs, agent a maintains a sourcing flow schedule (SFS), i.e., a list of agreed flow deliveries from $i, j \in P_a$. Each delivery $D_u = \{a_u, t_u, Q_u\}, D_u \in SFS$ is a vector of data that comprises: id of the predecessor sending flow (a_u), updated delivery time (t_u), and quantity of flow to be delivered (Q_u). Therefore, SFS is formally defined as

$$SFS = \{D_1, \ldots, D_u | t_x \le t_{x+1} \forall x \in [1, u]\}$$

As part of the collaboration between agent a and its sourcing agents $i \in P_a^1$, the latter provide a with the following information: (1) updated delivery time t_u estimates of pending deliveries $D_u \in SFS|a_u \in P_a^1$, upon request from a, and (2) notification of any disruption affecting delivery of pending $D_u \in SFS|a_u \in P_a^1$, initiated by $i \in P_a^1$. In order to eliminate incentives for untruthful status updates from $i \in P_a^1$, benefit collected by agents i from agent a must be monotonically decreasing with delivery time $t_{i \rightarrow a}$ and collected upon delivery of flow.

By combining updated D_u status information with flow consumption rate $\mu[S_a]$ estimations based on delivery needs to its successors S_a, agent a determines the nearest sourcing gap $SG = \{t_{SG}, Q_{SG}\}$ in current SFS, defining its time of occurrence t_{SG} and the required flow quantity Q_{SG}, as follows:

$$
t_{SG} = \min_{t' > t} \left\{ t' \,\middle|\, Q_t + \sum_{\substack{t_u \le t' \\ t_u \in D_u}} Q_u - (t' - t) \cdot \mu[S_a] = 0 \right\}
$$

$$
t_{u^*} = \min_{t_u \in D_u} \{ t_u | t_u > t_{SG} \}
$$

$$
Q_{SG} = \mu[S_a] \cdot (t_{u^*} - t_{SG})
$$

where Q_t is the quantity of flow in possession of agent a at time t and t_{u^*} is the time of the first delivery in SFS scheduled after t_{SG}. Sourcing gap is updated periodically, with frequency Δt_{SG}^{upd}, or after the notification of a disruption in $i \in P_a^1$.

Whenever the sourcing gap's time of occurrence t_{SG} is within the ordering horizon, i.e.,$t_{SG} \in [t, t + \Delta t_r]$, a sourcing or sharing protocol is triggered, depending on Q_{SG}. If the sourcing gap size is large enough not to exceed maximum overlap λ between any two contiguous deliveries D_u and D_{u+1}, i.e., $Q_{SG} \ge Q(1 - \lambda)$, then sourcing of Q units of flow from P_a^1 is requested (through sourcing protocol ST1P, described in Sect. 7.3.1). Otherwise, agent a attempts to cover the gap by sharing flow with one or more agents in P_a^2 (through sourcing protocol ST2P, described in Sect. 7.3.2).

After initiating ST1P or ST2P, agent a awaits protocol results, or a timeout condition $ST1P_{to}$ in the case of ST1P, and calculates the new sourcing gap SG^* in SFS. If the new sourcing gap is different from the gap that triggered ST1P or ST2P, i.e., $SG^* \ne SG$, agent a returns to the start of SFCP; else, it awaits for a timeout condition SG_{to} and repeats the sourcing requests through ST1P or ST2P, depending on Q_{SG}.

In summary, SFCP (Fig. 7.6) coordinates status information flow between agent a and its predecessors, and triggers specific sourcing protocols ST1P and ST2P. Protocol behavior can be adjusted by tuning six parameters described in Table 7.3.

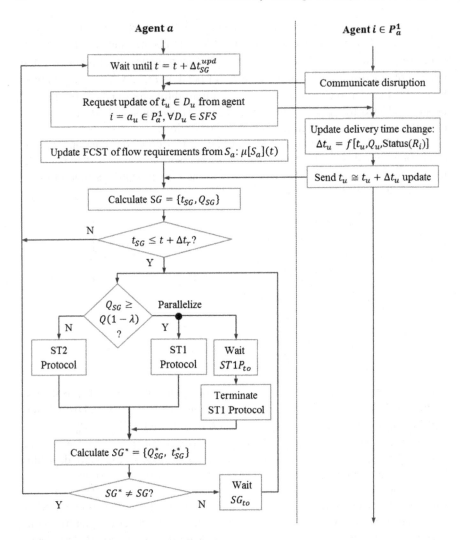

Fig. 7.6 Sourcing Flow Control Protocol (SFCP)

7.3.1 Sourcing from the Primary Team: ST1 Protocol

ST1 protocol (ST1P) is designed to leverage randomness in predecessors P_a^1 sourcing process, to their advantage and that of agent a. Through a bidding mechanism, each agent $i \in P_a^1$ can submit a sourcing proposal to a that best fits their internal resources' capacity and current status. Agent a initiates the bidding process by requesting bids for sourcing Q units if flow from agents $i \in P_a^1$. Then, each agent i determines the earliest feasible delivery time t_i they can send Q units of flow to agent a, and submit a bid $b_{i \to a} = \{t_{i \to a}, Q_{i \to a}\}$, $Q_{i \to a} = Q$.

Table 7.3 SFCP parameters and effects thereof

Parameter	Description	Effect on SFCP
Δt_{SG}^{upd}	Frequency of status updates from $i \in P_a^1$	• Controls agent a situation awareness • Lower values lead to higher situation awareness from a, at the expense of increased communication flow between a and $i \in P_a^1$
Δt_r	Look-ahead time for sourcing gap detection	• Controls the time distance between a disruption due to lack of flow and the start of ST1P or ST2P • Selection affects sourcing network design; higher values require smaller P_a^1 (for a given service level) but may increase the use of P_a^2 to cover SG
Q	Initial flow request quantity from a to its predecessors P_a	• Controls ordering frequency • Higher values increase exposure to disruptions in $i \in P_a$
λ	Maximum allowable overlap between any two contiguous deliveries D_h and D_{h+1} in SFS	• Controls flow redundancy (and,thus, availability of shareable flow in a) and ratio of ST1P/ST2P calls • Higher values increase: (1) protection against disruptions (requiring more storage resources) and (2) no. of ST1P vs. ST2P calls (T1 used more often)
$ST1P_{to}$	Time-out condition for ST1P execution	• Affects average no. of sourcing bids received from $i \in P_a^1$ when ST1P is called • Higher wait times allow for increased synergy between a flow requirements and $i \in P_a^1$ capacity
SG_{to}	Time-out condition for submitting a sourcing request for an unfulfilled SG	• Affects capacity of a to benefit from high availability in any agent $i \in P_a$ (quick deliveries may be rejected by λ; short wait times enable fast re-evaluation)

Instead of creating direct competition among predecessors for faster delivery, agent a will accept any sourcing bid that fits its own interests, regardless of the proposed delivery time. Upon receiving a bid, agent a evaluates whether to accept is or not, based on two conditions: (1) delivery time must be within scheduling horizon $t + \Delta t_{hzn}$, and (2) sum of overlap with current deliveries $D_u \in SFS$ must not exceed allowable overlap λ. Let λ_u be the overlap between $b_{i \rightarrow a}$ and D_u; then, λ_u is obtained as follows:

$$\lambda_u = \frac{\max\left\{0, \min_{x \in \{i \rightarrow a, u\}}\left\{t_x + \frac{Q_x}{\mu[S_a]}\right\} - \max_{x \in \{i \rightarrow a, u\}}\{t_x\}\right\}}{Q}$$

All bids that meet the acceptance criteria are further evaluated for quantity extension. If the gap between $b_{i \to a}$ and the delivery $D_{u^*} \in SFS$ that immediately follows the bid is within the allowable extension range $(0, \xi Q]$, then agent a requests a bid update to i for $Q = \mu[S_a](t_{u^*} - t_{i \to a})$. Following the update, bid $b_{i \to a}$ is converted to a delivery and added to SFS, and acceptance is communicated to agent i, which schedules the delivery on its internal control protocol.

The bid evaluation sequence described in the preceding paragraphs is repeated for every incoming bid, until a bid from every agent $i \in P_a^1$ is evaluated or ST1P is terminated by a time-out condition in SCFP.

ST1P fosters teaming among agents $\{a\} \cup P_a^1$ in the following manner:

(1) Agents $i \in P_a^1$ indirectly team up to source agent a, as the latter does not create competition among agents i.
(2) Agent a teams with each agent i by (possibly) accepting bids with long delivery times, to help agent i increase certainty regarding its successors' demand, even when congested or recovering from a disruption. In return, agent i provides accurate information when bidding.
(3) Agent a accepts overlap between deliveries from agents $i \in P_a^1$, which provides some protection against disruptions and enables agents i to bid for delivery times based on their own best interest.

In summary, ST1P (Fig. 7.7) controls sourcing from the primary by leveraging randomness and predecessor redundancy. Protocol behavior is controlled through three parameters presented in Table 7.4.

7.3.2 Sourcing from the Secondary Team: ST2 Protocol

ST2 protocol (ST2P) is designed to enable collaboration by flow sharing among agents in $\{a\} \cup P_a^2$. Teaming enables agents with excess flow to reduce the resources required to hold such flow and, simultaneously, aids agents with flow shortage to avoid disruptions.

When agent a detects a sourcing gap that cannot be covered by the primary team of predecessors P_a^1, it calls for bids among the community of agents that form the secondary sourcing team P_a^2. Each agent $j \in P_a^2$ defines a set of deliveries $b_{j \to a}^S$ that it can send to a within t_s, a delivery deadline selected by a. Each delivery proposal m specifies the flow quantity $Q_{j \to a,m}$ that can be shared at time $t_{j \to a,m}$.

$$b_{j \to a}^S = \left\{ (t_{j \to a,m}, Q_{j \to a,m}) | t_{j \to a,m} \leq t_s \right\}$$

$$t_s = t + \Delta t_r$$

After each agent $j \in P_a^2$ submits a sharing bid to a, the latter sorts delivery proposals based on $t_{j \to a,m}$ and accepts as many as needed to cover SG (or the largest fraction

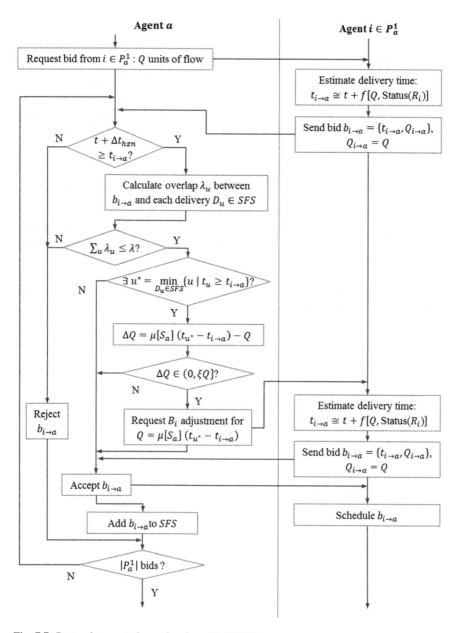

Fig. 7.7 Protocol to control sourcing from T1 (ST1P)

of *SG* possible). All accepted bids are scheduled into *SFS* and acceptance is communicated to corresponding agents $j \in P_a^2$, which schedule deliveries following their internal control protocols.

Table 7.4 ST1P parameters and effects thereof

Parameter	Description	Effect on ST1P
Δt_{hzn}	Scheduling horizon of agent a	Controls acceptance of bids with high $t_{i \to a}$ Higher values increase collaboration between a and i, as a is willing to accept longer delivery times due to congestion or disruptions in i
ξ	Maximum allowable extension of Q, on accepted bids $b_{i \to a}$	Affects how often sourcing gaps are closed by T2 Higher values leave fewer gaps, but increase exposure to disruptions in $i \in P_a^1$
λ	Maximum allowable overlap for adjacent deliveries in SFS (inherited from SFCP)	Affects acceptance ratio of bids $b_{i \to a}$; higher values increase acceptance ratio

The incentive of agents $j \in P_a^2$ to participate in sharing is two-fold: (1) reduction in resources used to hold excess flow, not needed to fulfill flow requests from S_j, and (2) possibility of receiving flow when needed to cover sourcing gaps. Fairness in the sharing process is ensured by making agent a responsible to (1) provide the resources needed to transfer flow from j to a and (2) supply resources to j equivalent to those spent sourcing the shared flow from its own predecessors P_j. Should

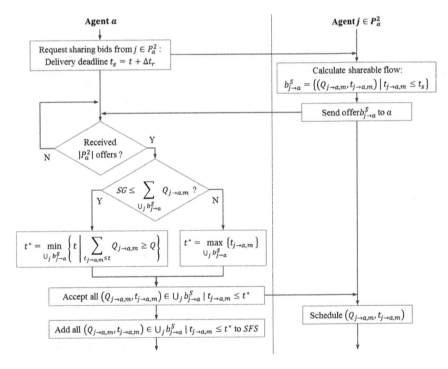

Fig. 7.8 Protocol to control sourcing from T2 (ST2P)

Table 7.5 ST2P parameter and effects thereof

Parameter	Description	Effect on ST2P
Δt_r	Time range for setting delivery deadline (inherited from SFCP)	Limits the look-ahead time for sharing proposals from $j \in P_a^2$ Higher values increase the possibility of receiving proposals from $j \in P_a^2$

resource transfer be infeasible for the SN under consideration, other methods to ensure sharing fairness are needed.

In summary, ST2P (Fig. 7.8) controls sourcing from the secondary team and leverages randomness and predecessor redundancy. Protocol behavior is controlled through three parameters presented in Table 7.5.

References

Al-Karaki, J. N., & Kamal, A. E. (2004). Routing techniques in wireless sensor networks: A survey. *IEEE Wireless Communications, 11*(6), 6–28. doi:10.1109/MWC.2004.1368893

Iyengar, S. S., Jayasimha, D. N., & Nadig, D. (1994). A versatile architecture for the distributed sensor integration problem. *IEEE Transactions on Computers, 43*(2), 175–185. doi:10.1109/12.262122

Jeong, W., & Nof, S. Y. (2009). A collaborative sensor network middleware for automated production systems. *Computers & Industrial Engineering, 57*(1), 106–113. doi:10.1016/j.cie.2008.11.007

Liu, Y., & Nof, S. Y. (2004). Distributed microflow sensor arrays and networks: Design of architectures and communication protocols. *International Journal of Production Research, 42*(15), 3101–3115. doi:10.1080/00207540410001699363

Nof, S. Y. (2007). Collaborative control theory for e-Work, e-Production, and e-Service. *Annual Reviews in Control, 31*(2), 281–292. doi:10.1016/j.arcontrol.2007.08.002

Qi, H., Iyengar, S. S., & Chakrabarty, K. (2001). Distributed sensor networks—a review of recent research. *Journal of the Franklin Institute, 338*(6), 655–668. doi:10.1016/S0016-0032(01)00026-6

Wan, J., Yuan, D., & Xu, X. (2008). A review of routing protocols in wireless sensor networks. In *Proceedings of the 4th International Conference on Wireless Communications, Networking and Mobile Computing* (pp. 1–4). Dalian, China. doi:10.1109/WiCom.2008.946

Wesson, R., Hayes-Roth, F., Burge, J. W., Stasz, C., & Sunshine, C. A. (1981). Network structures for distributed situation assessment. *IEEE Transactions on Systems, Man, and Cybernetics, 11*(1), 5–23. doi:10.1109/TSMC.1981.4308574

Chapter 8
Resilience by Teaming: Internal Resource Network Design, Flow Management, and Resource Control Protocols

Supply network agents transform flow received from their predecessors into output flow by executing a series of internal processes. These agents require a set of internal resources $r \in R_a$ which are oftentimes arranged in a network configuration under central control of agent a. Internal resources are classified in processes, i.e., resources that transform local input flow into local output flow, and storages, responsible for temporal stowage of flow to balance flow rate fluctuations.

Internal resource network resilience can be enabled by design and flow control protocol decisions. As found among reviewed articles in Sects. 4.3.1 and 4.3.2, resilience can be built-in by embedding excess capacity and storage. However, these strategies entail one-time costs, e.g., equipment purchase and installation, and operation costs, e.g., energy consumption and maintenance, and need to be utilized efficiently to increase resilience in a sustainable manner. To this end, the principle of Fault-tolerance by Teaming from CCT lays the foundation to leverage weaker, thus less costly, resources to achieve a performance equal to, or better than, that achievable by a single flawless, and costlier, resource.

Team formation in internal resource networks (IRNs) is influenced by parameters selected at the design stage and decisions made in real-time based on available status information. Design choices define limits to the ability of flow control protocols to re-parameterize an IRN in order to protect resources at risk of disruptions. The following sections present design guidelines and a flow control protocol to achieve resilient operation of internal resource networks.

8.1 Design of Internal Resource Networks for Resilience Through Resource Teaming

IRNs for different applications may differ in number of transformation stages and relative sequence thereof. Furthermore, designs for a given application may present various topologies. These, despite being partially defined by the nature of the

© Springer International Publishing AG 2018
R. Reyes Levalle, *Resilience by Teaming in Supply Chains and Networks*,
Automation, Collaboration, & E-Services 5, DOI 10.1007/978-3-319-58323-5_8

transformations required to convert agent's input flow into output flow, are affected by capacity sizing decisions. Addressing the general design of IRNs exceeds the scope of this book; nevertheless, the basic requirements for designing resilient IRNs through teaming are outlined in the next sections.

8.1.1 Resource Parallelism: Effect on Throughput Average and Variability

Magazine and Stecke (1996) observed that production lines with parallel processes show higher average throughput that those without resource parallelism. From these results, the authors inferred that arranging resources in parallel provides some level of protection against blocking, i.e., process idling due to lack of storage space, and starvation, i.e., process idling due to lack of input flow. Nevertheless, no formal description of the phenomena is provided.

Blocking and starvation can manifest on IRNs in different levels, depending on IRN topology, configuration, and operation parameters. These negative effects are triggered by variability; a balanced IRN with deterministic processes, i.e., without variability, will not be subject to negative interactions affecting throughput. Hence, the effect observed by Magazine and Stecke (1996) can be explained by a reduction on processing stages' variance, obtained from arranging resources in parallel.

Consider a processing stage (Fig. 8.1) with n equal processes in parallel and total capacity ϑ. Let $p(r^P)$ be the probability that process r^P is available to process flow during interval Δt. Then, assuming the processes are independent, the processing stage expected throughput is

$$E[TP] = \sum_{q=0}^{n} \frac{q\vartheta}{n} \binom{n}{q} [1 - p(r^P)]^{n-q} [p(r^P)]^q = \frac{\vartheta}{n} \sum_{q=0}^{n} q \binom{n}{q} [1 - p(r^P)]^{n-q} [p(r^P)]^q$$

$$= \frac{\vartheta}{n} E[Bin(n, p(r^P))] = \frac{\vartheta}{n} np(r^P) = \vartheta p(r^P)$$

Fig. 8.1 Parallel processing in IRNs

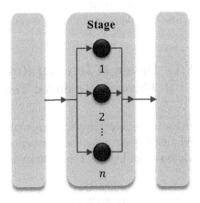

Similarly, the variance of the processing stage's throughput can be obtained from

$$Var[TP] = E[TP^2] - E[TP]^2 = E[TP^2] - [\vartheta p(r^P)]^2$$

where

$$E[TP^2] = \sum_{q=0}^{n} \left(\frac{q\vartheta}{n}\right)^2 \binom{n}{q} [1 - p(r^P)]^{n-q} [p(r^P)]^q = \frac{\vartheta^2}{n^2} \sum_{q=0}^{n} q^2 \binom{n}{q} [1 - p(r^P)]^{n-q} [p(r^P)]^q$$

$$= \frac{\vartheta^2}{n^2} \left[Var[Bin(n, p(r^P))] + E[Bin(n, p(r^P))]^2 \right] = \frac{\vartheta^2}{n^2} \left[np(r^P) [1 - p(r^P)] + [np(r^P)]^2 \right]$$

Therefore,

$$Var[TP] = \frac{\vartheta^2 p(r^P)[1 - p(r^P)]}{n}$$

Clearly, $Var[TP]$ is monotonically decreasing with the number of parallel processes n.

The previous result sets the conditions required to benefit from teaming weaker resources to achieve lower variability. Consider two alternative stage designs $s = 1, 2$ capable of yielding the same throughput $E[TP]$, with capacities ϑ_s, number of parallel processes n_s, and probability of a process being available $p_s(r^P)$—where $p_2(r^P) < p_1(r^P)$. Then,

$$E[TP_1] = \vartheta_1 p_1(r^P) = E[TP_2] = \vartheta_2 p_2(r^P) \rightarrow \vartheta_2 = \vartheta_1 \frac{p_1(r^P)}{p_2(r^P)} > \vartheta_1$$

$$\frac{Var[TP_1]}{Var[TP_2]} = \frac{n_2 \vartheta_1^2 p_1(r^P)[1 - p_1(r^P)]}{n_1 \vartheta_2^2 p_2(r^P)[1 - p_2(r^P)]} = \frac{n_2 p_2(r^P)[1 - p_1(r^P)]}{n_1 p_1(r^P)[1 - p_2(r^P)]}$$

In order for $s = 2$, i.e., the alternative comprising weaker resources, to have lower variability than $s = 1$, the following must hold

$$\frac{Var[TP_1]}{Var[TP_2]} > 1 \rightarrow \frac{n_2}{n_1} > \frac{p_1(r^P)[1 - p_2(r^P)]}{p_2(r^P)[1 - p_1(r^P)]}$$

Figure 8.2 (left) shows the minimum ratio n_2/n_1 required to obtain lower variability, hence higher resilience, through teaming, for alternative combinations of $p_1(r^P)$ and $p_2(r^P)$. Figure 8.2 (right) presents the ratio between stage capacities ϑ_2/ϑ_1 for alternative combinations of $p_1(r^P)$ and $p_2(r^P)$. For example, consider a process with $n_1 = 2$ and $p_1(r^P) = 0.9$. A stage with $n_2 = 12$, $p_2(r^P) = 0.6$ and $\vartheta_2 = 1.5 \vartheta_1$ will have lower variance than configuration 1 with the same expected throughput.

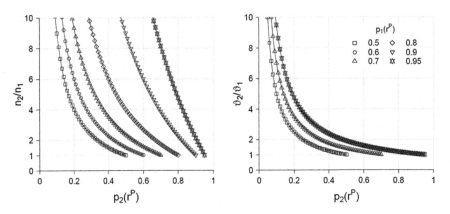

Fig. 8.2 Ratio of parallel processes and capacities for reduced stage variability through teaming

8.1.2 Storage Design Guidelines for Management Under Teaming Protocols

Location and capacity sizing of storage in IRNs is an NP-hard problem that has been extensively researched, as evidenced by the literature reviews in Gershwin and Schor (2000) and Demir et al. (2012). Storage provides de-coupling among interconnected processes in an IRN, allowing the latter to operate with less negative interactions due to blocking or starvation. Therefore, larger storage capacity leads to higher average throughput (Meester and Shanthikumar 1990; Magazine and Stecke 1996).

From a resilience perspective, IRNs should be capable of providing stable output flow, regardless of input flow fluctuations and/or internal resource disruptions. Despite the vast amount of articles addressing the effect of storage capacity on IRN throughput, minor attention has been given to the effect storage capacity has on throughput variability. Recently, Assaf et al. (2013) concluded that the effects of storage capacity on throughput variability are highly dependent on system characteristics and configuration. However, for any IRN configuration, variability converges to that of the bottleneck process for large storage capacities (Gershwin and Schor 2000; Assaf et al. 2013).

Optimal storage capacity decisions depend on the characteristics of the processes requiring protection. Blumenfeld and Li (2005) show effectiveness of storage to provide protection against failures depends not only on the size thereof, but also on IRN throughput and downtime duration of the processes immediately before and/or after the storage. In this line, experiments in Gershwin and Schor (2000) show that increased downtime duration reduces protection effectiveness to a greater extent than shorter downtime frequencies.

Analyses of storage effectiveness to provide protection against disruptions are based on simple management rules—e.g., approaches presented in Sect. 5.2—which usually fail to leverage prognostic capabilities and process-storage

Fig. 8.3 Storage and process
interaction: blocking and
starvation versus design
characteristics

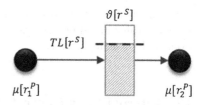

collaboration. Next, storage design based on active resource teaming is discussed to provide guidelines to create IRNs capable of higher resilience via teaming protocols.

Consider a storage r^S between two processes r_1^P and r_2^P to provide de-coupling and avoid negative interaction (Fig. 8.3) and let r^S have capacity $\vartheta[r^S]$ and target level $TL[r^S]$. Let $t_{down}\left[r_x^P\right]$ and $\mu\left[r_x^P\right]$ be the distribution of downtime duration and the flow rate of process x, respectively, and assume that (1) the time required for r_1^P to replenish storage level to $TL[r^S]$ is lower than the minimum time between disruptions in r_2^P and (2) the time required for r_2^P to reduce storage level to $TL[r^S]$ is lower than the minimum time between disruptions in r_1^P.

Then, the probability of preventing blocking p_{NB} and starvation p_{NS} are

$$p_{NB}\left(r_1^P\right) = P\left[t_{down}\left[r_2^P\right] \leq \frac{\vartheta[r^S] - TL[r^S]}{\mu\left[r_1^P\right]}\right]$$

$$p_{NS}\left(r_2^P\right) = P\left[t_{down}\left[r_1^P\right] \leq \frac{TL[r^S]}{\mu\left[r_2^P\right]}\right]$$

Under the aforementioned conditions, it is possible to dynamically adjust protection against blocking or disruption by selecting an adequate $TL[r^S]$. To this end, processes must collaborate to forecast whether a disruption in r_1^P or r_2^P is more likely to occur next and form a team in order to adjust $TL[r^S]$ accordingly. Assuming processes can accurately predict blocking and starvation potential, the maximum levels of protection achievable by adjusting $TL[r^S]$ are

$$\max_{TL[r^S] \leq \vartheta[r^S]} \left\{p_{NB}\left(r_1^P\right)\right\} = P\left[t_{down}\left[r_2^P\right] \leq \frac{\vartheta[r^S]}{\mu\left[r_1^P\right]}\right], \quad TL[r^S] = 0$$

$$\max_{TL[r^S] \leq \vartheta[r^S]} \left\{p_{NS}\left(r_2^P\right)\right\} = P\left[t_{down}\left[r_1^P\right] \leq \frac{\vartheta[r^S]}{\mu\left[r_2^P\right]}\right], \quad TL[r^S] = \vartheta[r^S]$$

Based on the former design equations, it is possible to define $\vartheta[r^S]$ as the minimum value that simultaneously satisfies the blocking and starvation protection levels under teaming protocols.

8.2 Internal Flow Control Protocol (IFCP)

Agent a must coordinate its internal resource network in order to transform input flow into output flow while meeting QoS requirements of successors S_a. The Internal Flow Control Protocol (IFCP) dynamically updates SLAs between agent a and its internal resources $r_n \in R_a$ in order to minimize long-term QoS losses and flow costs by anticipating, and minimizing the effect of, disruptions. IFCP relies on four main components: an IRN designed to enable process-process and process-storage teaming, an early conflict detection mechanism, a database containing optimized configurations for each IRN state, and rules to team resources to anticipate and overcome disruptions.

Consider an IRN with $R_a = \{r_1, \ldots, r_n\}$ resources that transform agent a's input flow in output flow. Flow transformation occurs continuously; however, as resources transition through different status, output flow may be affected by the combination of resource states at a given point in time. Let $\Phi[r_n]$ be the set of possible states $\phi[r_n]_m$ for resource $r_n \in R_a$ and $\phi[r_n](t)$ the state of r_n at time t. Then, the state of an IRN at time t is fully described by

$$\phi[R_a](t) = \{\phi[r_1](t), \ldots, \phi[r_n](t)\}$$

Assuming the IRN state is permanent, it is possible to obtain a unique resource configuration in order to ensure maximum fulfillment of QoS requirements for agent a. Let $QoS_{a \to j}^{SLA}$ be the target QoS required by agent j under $SLA_{a \to j}$, then IRN configuration must be optimized to minimize QoS losses as follows:

$$\min \sum_{j \in S_a} QoS_{a \to j}^{SLA} - QoS_{a \to j}$$

$$\text{s.t.} : \quad QoS_{a \to j}^{SLA} \geq QoS_{a \to j}$$

$$\phi[r_n] = \phi[r_n](t) \quad \forall r_n \in R_a, \quad \phi[r_n](t) \in \phi[R_a](t)$$

Optimal configuration for each $\phi[R_a]$ is stored in a configuration database (IRN-CDB, Internal Resource Network—Configuration Data Base) to be later retrieved by the Internal Flow Control Protocol.

Using real-time information for internal resource control enables dynamic adjustment to changing system conditions; however, the ability to anticipate disruptions requires not only real-time resource status information but also analysis and prognosis capabilities to enable control protocols to forecast likely system evolution. Through an application-specific Early Conflict Detection Tool (ECDT), historical resource performance and current status information are integrated to select and allocate the optimal combination of collaborative protection measures, e.g., increase/reduce storage target levels or parallelize flow, to prepare for a possible disruption and/or remediate one after its occurrence.

Figure 8.4 presents the Internal Flow Control Protocol. Agent a initiates an evaluation of IRN re-configuration needs after a resource communicates a change in state or with frequency Δt_{ECDT}^{upd}. When IFCP is triggered, agent a requests all resources $r \in R_a$ to send a state update. Based on individual IRN state $\phi[R_a]$, agent

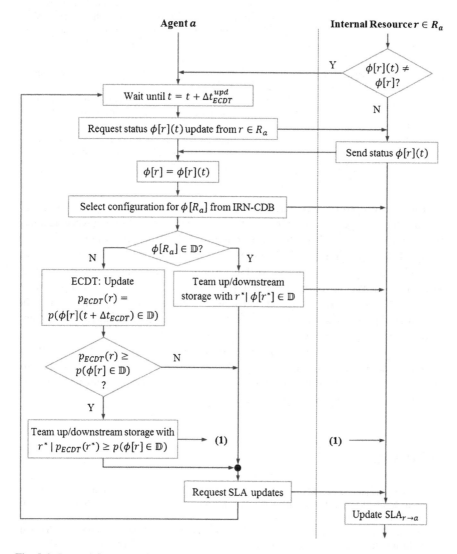

Fig. 8.4 Internal flow control protocol (IFCP)

Table 8.1 IFCP parameters and effects thereof

Parameter	Description	Effect on IFCP
Δt_{ECDT}^{upd}	Frequency of status updates from $r \in R_a$	• Controls agent a situation awareness, relative to its internal resources • Lower values lead to higher situation awareness from a, at the expense of increased communication flow between a and $r \in R_a$
Δt_{ECDT}	Look-ahead horizon for ECDT prognostics	• Controls available time to prepare for (possible) disruptions • Higher values enable earlier implementation of preparedness measures, resulting in higher protection at the expense of higher average WIP • Lower values avoid large increases in average WIP levels and (possibly) increase ECDT accuracy
$p(\phi[r] \in \mathbb{D})$	Preparedness measures threshold for resource r	• Affects the number of times protective measures are taken • Higher values (possibly) reduce the number of false positives, increasing effectiveness of protective measures. • Higher values (possibly) increase the number of false negatives, resulting in lack of protection against disruptions

a retrieves the optimal configuration from IRN-CDB. In the case $\phi[R_a]$ is a disrupted state $\mathbb{D}$, agent a combines optimal configuration with recovery teaming measures to overcome disruption at resource/s r^*. If IRN state is normal, i.e., not disrupted, agent a applies ECDT within a prognostics horizon Δt_{ECDT} to assess the probability $p(\phi[r](t + \Delta t_{ECDT}) \in \mathbb{D})$ that resource r is disrupted within the prognostic horizon. Whenever the disruption probability exceeds a predefined parameter $p(\phi[r] \in \mathbb{D})$, anticipatory protective teaming measures are combined with the optimal configuration to defend resources at risk. SLAs updates are communicated to resources $r \in R_a$, therefore updating IRN configuration.

In summary, IFCP (Fig. 8.4) controls IRN configuration to leverage available storage capacity/levels and resource parallelism to achieve long-term flow stability and ensure maximum fulfillment of $SLA_{a \to j}$. Protocol behavior is controlled through three parameters presented in Table 8.1.

References

Assaf, R., Colledani, M., & Matta, A. (2013). Analytical evaluation of the output variability in production systems with general Markovian structure. *OR Spectrum, 36*(3), 799–835. doi:10.1007/s00291-013-0343-6

Blumenfeld, D. E., & Li, J. (2005). An analytical formula for throughput of a production line with identical stations and random failures. *Mathematical Problems in Engineering, 2005*(3), 293–308. doi:10.1155/MPE.2005.293

Demir, L., Tunali, S., & Eliiyi, D. T. (2012). The state of the art on buffer allocation problem: A comprehensive survey. *Journal of Intelligent Manufacturing, 25*(3), 371–392. doi:10.1007/s10845-012-0687-9

Gershwin, S. B., & Schor, J. E. (2000). Efficient algorithms for buffer space allocation. *Annals of Operations Research, 93*(1–4), 117–144. doi:10.1023/A:1018988226612

Magazine, M. J., & Stecke, K. E. (1996). Throughput for production lines with serial work stations and parallel service facilities. *Performance Evaluation, 25*(3), 211–232. doi:10.1016/0166-5316(95)00005-4

Meester, L. E., & Shanthikumar, J. G. (1990). Concavity of the throughput of tandem queueing systems with finite buffer storage space. *Advances in Applied Probability, 22*(3), 764–767. doi:10.2307/1427472

Chapter 9
Resilience by Teaming: Distribution Network Design and Flow Management Protocols

Agents in a supply network deliver flow to other agents in the network or to entities outside the scope of the SN. Flow distribution occurs through flow links $fl_{i \to j}$ which are subject to disruptions that affect service levels, possibly extending far beyond the reach of a single link. Therefore, creating resilient links and managing flow to avoid and overcome disruptions without impact on the service level is of utmost importance.

This chapter presents two protocols from the Resilience by Teaming framework: Distribution Network Formation/Re-configuration Protocol (DNF/RP) and Distribution Flow Control Protocol (DFCP). DNF/RP implements a fuzzy network formation and reconfiguration strategy designed to create routes that balance the cost of flow distribution with the ability to re-configure the route in the face of disruptions. DFCP performs real-time monitoring and route re-configuration to avoid disruptions and ensure flow delivery according to the predefined SLAs. The use of resilient distribution protocols in combination with the sourcing and internal control protocols introduced in preceding chapters provides the foundation for Resilience by Teaming in supply networks.

9.1 Distribution Dynamics in Supply Networks

Delivery of flow between agents takes place through flow links $fl_{i \to j}$, which form whenever agent i agrees to send flow to agent j under a pre-defined SLA. Hitherto, these abstract constructions have been treated as a direct connection between i and j. Nevertheless, delivery of flow from one agent to another oftentimes requires a set of

© Springer International Publishing AG 2018
R. Reyes Levalle, *Resilience by Teaming in Supply Chains and Networks*,
Automation, Collaboration, & E-Services 5, DOI 10.1007/978-3-319-58323-5_9

Fig. 9.1 Intermediary network in a flow link

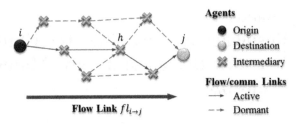

intermediary agents that collaborate to create link $fl_{i \to j}$ (Fig. 9.1). Collaboration among these parties is overseen by the origin agent (i) or the destination agent (j).

Consider a set $H_{i \to j}$ of intermediary agents h which may connect to other intermediaries in $H_{i \to j}$, the origin (i) agent, and/or destination (j) agent in order to send flow. Dormant links, i.e., connections not being used to transmit flow or messages, must form a network containing at least one path $\delta_{i \to j}$ between origin and destination, in order for $fl_{i \to j}$ to exist.

Intermediary agents h receive flow transmission requests from several i, j pairs simultaneously and require a processing time $\mu[h]^{-1}$ to execute each request. In general, $\mu[h]$ will depend on agent h's load and status. Higher loads lead to congestion and, consequently, lower $\mu[h]$. Similarly, disruptions temporarily prevent intermediaries from transferring flow, increasing $\mu[h]^{-1}$. Flow between any two agents in $\{i, j\} \cup H_{i \to j}$ is also subject to congestion. Flow time $ft_{h \to h'}$, with $h, h' \in \{i, j\} \cup H_{i \to j}$ depends on the amount of traffic between h and h', as well as the characteristics of the connection.

Without congestion and disruption, choice of an optimal path to realize $fl_{i \to j}$ would be straightforward: select the minimum cost path $\delta_{i \to j}$ capable of delivering flow to j within a predefined due date $t_{i \to j}$. However, given the dynamic conditions of agents in $H_{i \to j}$ and their interconnections, path selection requires protocols capable of enabling resilience by leveraging path redundancy. To this end, Sections 0 and 0 present a network formation/re-configuration protocol to dynamically select flexible paths $\delta_{i \to j}$ and a flow control protocol to enable disruption/congestion avoidance.

9.2 Distribution Network Formation/Re-configuration Protocol (DNF/RP)

Consider a SN agent a responsible for delivering flow to a set of successors S_a through flow links $fl_{a \to j}, j \in S_a$. Each flow link has an associated set of intermediaries $H_{a \to j}$ whose interconnections define a set $\Delta_{a \to j}$ of paths $\delta_{a \to j}$ between a and j. $H_{a \to j}$

may change over time, by incorporating or eliminating intermediaries; however, for any instance of $H_{a \to j}$ it is possible to find the associated $\Delta_{a \to j}$ using an off-line depth-first search (Migliore et al. 1990).

The collection of all sets of intermediaries $H_{a \to j}$ associated to flow links $fl_{a \to j}$ determine a delivery network that connects agent a to $j \in S_a$. At any point in time, a fraction of the network is active transmitting flow while the rest remains dormant. Intermediaries are activated or made dormant dynamically by the application of Distribution Network Formation/Re-configuration Protocol (DNF/RP), shown in Fig. 9.3.

In DNF/RP, agent i selects a team of intermediaries $h \in H_{i \to j}$ to deliver flow to agent j by evaluating their path performance in terms of cost, flexibility, and margin left for contingency measures. Path origin i can be agent a, on the initial path decision, or any agent $h \in H_{a \to j}$, when path re-configuration is required. In any case, agent a is responsible for selecting DNF/RP parameters, as a is accountable to j for flow delivery within pre-arranged SLA.

DNF/RP begins when agent i retrieves all feasible paths $\Delta_{i \to j} \subseteq \Delta_{a \to j}$ and requests intermediaries $h \in \Delta_{i \to j}$ to estimate processing rate $\mu[h]$ and flow time $ft_{h \to h'}$, $h' \in \Delta_{i \to j}$. Based on these estimations, the cost ω_h of processing flow at intermediary h, and the cost $\omega_{h \to h'}$ of sending flow from h to h', agent i determines the following measures for each path $\delta_{i \to j}$:

$$Slack\left[\delta_{i \to j}\right] = t_{a \to j} - \left[t + \sum_{(h \to h') \in \delta_{i \to j}} ft_{h \to h'} + \sum_{h \in \delta_{i \to j}} \mu[h]^{-1} \right]$$

$$Cost\left[\delta_{i \to j}\right] = \sum_{(h \to h^{e'}) \in \delta_{i \to j}} \omega_{h \to h'} - \sum_{h \in \delta_{i \to j}} \omega_h$$

$$Flex\left[\delta_{i \to j}\right] = \sum_{h \in \delta_{i \to j}} \left[deg^{OUT}(h) - 1 \right] \left[d\left(\delta_{h \to j}\right) \right]^{\kappa}$$

$Slack\left[\delta_{i \to j}\right]$, the time difference between agreed delivery time $t_{a \to j}$ and expected delivery time under path $\delta_{i \to j}$, quantifies the margin of time available to take contingency measures while flow moves from i to j. Paths with higher margins give

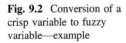

Fig. 9.2 Conversion of a crisp variable to fuzzy variable—example

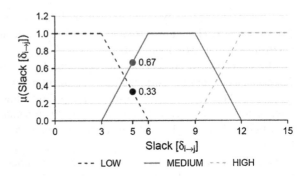

more protection against unforeseen congestion and/or disruptions. $Cost[\delta_{i\to j}]$ measures the total cost to deliver flow through $\delta_{i\to j}$. $Flex[\delta_{i\to j}]$ quantifies the available flexibility of path $\delta_{i\to j}$, measured by the number of alternative connections each intermediary h has, weighed by a function of the distance to the destination j. Then, paths that contain intermediaries with high out-degrees and/or give agent a an earlier the opportunity to revise any decision made have a higher flexibility coefficient.

In order for agent a or any intermediary h to select the best path based on the abovementioned criteria, agent a defines a set of fuzzy logic scoring rules FR_a to evaluate each path $\delta_{i\to j}$.

$$\text{IF } Slack[\delta_{i\to j}] \text{ IS } X \text{ AND } Cost[\delta_{i\to j}] \text{ IS } Y \text{ AND } Flex[\delta_{i\to j}] \text{ IS } Z \text{ THEN } Score[X, Y, Z]$$

Categories, e.g., low, medium, high, are defined for each measure and a fuzzy score is set for each combination of measure categories. Then, $Slack[\delta_{i\to j}]$, $Cost[\delta_{i\to j}]$, and $Flex[\delta_{i\to j}]$ are converted to fuzzy values based on the fuzzy variables defined by a. For instance, consider the fuzzy variable for $Slack[\delta_{i\to j}]$ shown in Fig. 9.2. A crisp value of 5 hours would belong to a classification of LOW with degree 0.33 and to the classification of MEDIUM with degree 0.67, based on the scales shown. The score obtained from each rule is then aggregated (with a fuzzy sum) and de-fuzzified, to obtain a crisp value for $\delta_{i\to j}$. The process is repeated for all paths in $\Delta_{i\to j}$ and the best-scoring path is selected.

Although not addressed in DNF/RP, the set of fuzzy scoring rules can be dynamically updated, i.e., by creating/eliminating, and/or weighing rules, using a learning classifier system (Sigaud and Wilson 2007).

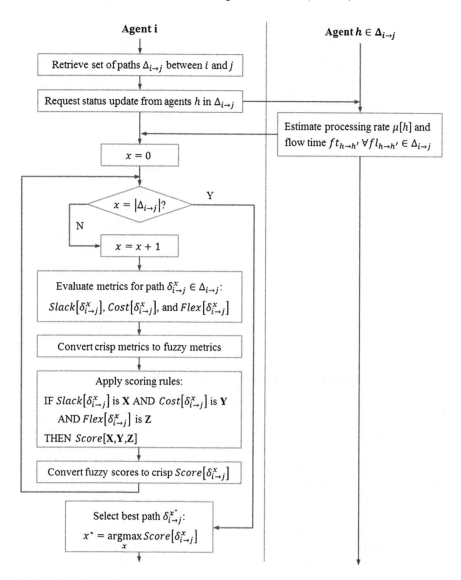

Fig. 9.3 Distribution Network Formation/Re-configuration Protocol (DNF/RP)

In summary, DNF/RP (Fig. 9.3) controls path formation from agent i to agent j, to realize flow links $fl_{a \to j}$. Protocol behavior is controlled through three parameters presented in Table 9.1.

Table 9.1 DNF/RP parameters and effects thereof

Parameter	Description	Effect on DNF/RP
FR_a	Fuzzy variables for $Slack[\delta_{i\rightarrow j}]$, $Cost[\delta_{i\rightarrow j}]$, and $Flex[\delta_{i\rightarrow j}]$; score for each combination of fuzzy values	• Affect protocol ability to produce good solutions and discern among various levels in evaluation metrics • Fuzzy variables with lower number of values have lower screening ability which may lead to inability to choose among extremely good (or bad) paths • Larger number of values in fuzzy variables lead to more a more complex set of rules by increasing the number of combinations
κ	Parameter weighing distance to destination in $Flex[\delta_{i\rightarrow j}]$	• Controls the relative importance towards path flexibility of the position of intermediary h within path $\delta_{i\rightarrow j}$ • Higher values make distance to destination more relevant; intermediaries closer to agent i allow adjusting decisions earlier, thus conferring higher flexibility

9.3 Distribution Flow Control Protocol (DFCP)

Flow control comprises two processes: (1) definition of initial path $\delta_{a\rightarrow j}$ and (2) re-evaluation of current path $\delta_{i\rightarrow j}$ at intermediaries i. Initial path definition is performed by agent a through DNF/RP. Following the selection of a path $\delta_{a\rightarrow j}$, agent a communicates the routing $\delta_{h\rightarrow j} \in \delta_{a\rightarrow j}$ to each intermediary h. Additionally, agent a sends its intermediaries in $\delta_{a\rightarrow j}$ three parameters required for process (2): the requested delivery time ($t_{a\rightarrow j}$), the slack value below which path re-evaluations are always required (ρ_a); and the set of fuzzy scoring rules to update delivery path (FR_a). Finally, agent a executes the first section of $\delta_{a\rightarrow j}$, namely $a \rightarrow h$.

Initial conditions under which path $\delta_{a\rightarrow j}$ was selected may change over time, possibly requiring a path re-configuration. Upon receiving flow, an intermediary i requests an status update from other intermediaries in $\delta_{i\rightarrow j}$ and re-evaluates $Slack[\delta_{i\rightarrow j}]$. If enough slack is available, the flow is routed to the next destination through $i \rightarrow h \in \delta_{i\rightarrow j}$. Should $Slack[\delta_{i\rightarrow j}]$ be under ρ_a, agent i calls DNF/RP and defines the best possible path $\delta_{i\rightarrow j}^{x^*}$ given the current conditions in $\Delta_{i\rightarrow j} \subseteq \Delta_{a\rightarrow j}$. Then, results are communicated to intermediaries in $\delta_{i\rightarrow j}^{x^*}$, defined by a. Finally, flow is routed to the next destination through $i \rightarrow h \in \delta_{i\rightarrow j}^{x^*}$.

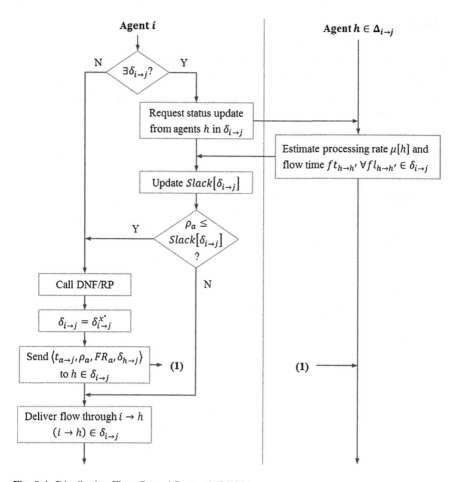

Fig. 9.4 Distribution Flow Control Protocol (DFCP)

Table 9.2 DFCP parameters and effects thereof

Parameter	Description	Effect on DFCP
ρ_a	Slack tolerance of agent a	• Controls the threshold for re-evaluation of delivery path • Higher values lead to re-evaluation with higher time margins; thus, increasing the protocol's anticipatory capabilities

The process described in the preceding paragraphs constitutes the Distribution Flow Control Protocol, DFCP, presented in Fig. 9.4. Protocol behavior is controlled through a parameter presented in Table 9.2.

References

Migliore, M., Martorana, V., & Sciortino, F. (1990). An algorithm to find all paths between two nodes in a graph. *Journal of Computational Physics, 87*(1), 231–236. doi:10.1016/0021-9991 (90)90235-S

Sigaud, O., & Wilson, S. W. (2007). Learning classifier systems: A survey. *Soft Computing, 11*(11), 1065–1078. doi:10.1007/s00500-007-0164-0

Chapter 10
Case Study A: Internal Flow Control Protocol Applied to Unreliable Production Lines

Production lines can be described as IRNs of a supply network agent, consisting of interconnected processes and storages that transform raw materials into finished products. In almost every case, their objective involves maintaining constant throughput to match demand with minimum WIP. Nonetheless, resources are subject to random failures, creating local downtimes (i.e., at resource level) that can spread through the IRN affecting other non-failed resources and, ultimately, the QoS an agent provides to its successors in the SN. As pointed out by Hudson et al. (2014), the problem of unbalanced production lines with unreliable equipment has received significant attention for over half a century. Nevertheless, despite advances in modeling and control of these systems, modern high speed production systems are still challenged by the succession of process failures that lead to line downtimes and, consequently, throughput reduction. This problem is becoming ubiquitous, as production systems become "leaner" and are increasingly required to process a more diverse mix of products with higher speeds and performance.

Case Study A analyzes the relative merits of a teaming-based flow control protocol, IFCP (Reyes Levalle et al. 2013; Reyes Levalle 2015), versus traditional production line control methods such as Kanban, CONWIP, DWIP, and Base-stock. Results show that a *teaming* approach to resilient production line control is capable of reducing work-in-process and throughput variability when compared to other traditional control approaches for the same line throughput performance.

10.1 Introduction

Most research conducted in the last decades has focused on storage management techniques to address the problem of production line control under uncertain machine behavior, in order to reduce the WIP required to achieve a given

R. Reyes Levalle, *Resilience by Teaming in Supply Chains and Networks*,
Automation, Collaboration, & E-Services 5, DOI 10.1007/978-3-319-58323-5_10

throughput level. Storage management techniques (reviewed in Sect. 5.2) consist in control loops that link storages and processes so that job release is regulated by a system state variable. Most approaches are passive, i.e., there is no adaptive/ predictive change in logic based on current/prognosed network status.

However, recently, research has extended to some degree into active control techniques such as DWIP (Yang et al. 2006) and dynamic control through learning agents (Paternina-Arboleda and Das 2001); also integrating production machines parameters into the control loop, e.g., optimal controller selection (Ma and Koren 2004), simulation based real-time decision making (Dalal et al. 2003). Despite some progress towards active control protocols to deal with intrinsic process variability, high volume automated production systems are still challenged by the succession of process failures that lead to line downtimes and, consequently, throughput reduction. Whenever a process fails, WIP inventory builds up and, eventually, another machine stops due to lack of storage space. The extent to which this dynamic, emergent, negative interaction among internal agent resources impacts its ability to maintain QoS depends on the structure of the IRN and the control protocols managing flow and line configuration.

10.1.1   Production Network Formalism

Consider a production network where input flow, received from an agent $i \in P_a$, is processed through a set of resources $r \in R_a$ to produce output flow that is later sent to agents $j \in S_a$. Figure 10.1 shows the formal description of the production resources, i.e., processes and storages, based on the SN formalism in Sect. 2.3. In general, a resource r receives flow from resource r' in a production network through a flow link $fl_{r' \to r}$ and transforms such flow into output according to a set of deterministic or stochastic transformation functions $f_r^{I/O}$. These functions enable resources to process one or several flow streams, resulting in single or multi-flow (product) production lines, respectively. Moreover, variations in $f_r^{I/O}$ enable the formalism to account for heterogeneous production processes. In the case of storages, flow is not transformed but stowed, with capacity $\vartheta[r]$.

Resources can actively modify their behavior through a control logic which acts, for instance, on job release rates, material acceptance into storage, and parameters

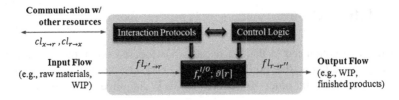

Fig. 10.1 Production line resource formalism

in $f_r^{I/O}$. Moreover, each resource $r \in R_a$ interacts with other resources in R_a and agent a through a set of interaction policies that regulate communication and information exchange via communication links $cl_{r \to x}$ and $cl_{x \to r}$ where $x \in R_a \cup \{a\}$.

Processes may experience failures, defined as any event deviating performance from normal parameters. Following the definitions in Sect. 8.2, the state of process r_n^P can be defined by comparing its instantaneous capacity $\vartheta[r_n^P](t)$ to its design capacity $\vartheta[r_n^P]$; whenever the former is lower than the latter, the process is failed. Note that failed does not necessarily implies $\mu[r_n^P] = 0$, as instantaneous capacity can be decreased in any continuous amount between $(0, \vartheta[r_n^P]]$.

$$\phi[r_n^P](t) = \begin{cases} 1 & \text{if } \vartheta[r_n^P](t) = \vartheta[r_n^P] \\ 0 & \text{if } \vartheta[r_n^P](t) < \vartheta[r_n^P] \end{cases}$$

Similarly, storage status can be defined by comparing instantaneous capacity vs. target level. Let $\varepsilon[r_m^S]$ be the tolerance around target level $TL[r_m^S]$ to consider storage r_m^S within target range; then,

$$\phi[r_m^S](t) = \begin{cases} 1 & \text{if } |\vartheta[r_m^S](t) - TL[r_m^S]| \leq \varepsilon[r_m^S] \\ 0 & \text{if } |\vartheta[r_m^S](t) - TL[r_m^S]| > \varepsilon[r_m^S] \end{cases}$$

Based on the previous definitions, the state at time t of a production network with n processes and m storages can be fully described by

$$\phi[R_a](t) = \{\phi[r_1^P](t), \dots, \phi[r_n^P](t), \dots, \phi[r_1^S](t), \dots, \phi[r_m^S](t)\}$$

and the state space complexity of the IRN is 2^{n+m}.

10.2 Collaborative Production Control in a Tissue Paper Production Network

In order to apply IFCP to a real-world production network, three components need to be tailored to such application: IRN-CDB, the database of optimal IRN configurations for each state, ECDT, a tool/mechanism to forecast likely-to-occur failures in the near future, and teaming rules to anticipate and respond to disruption.

IFCP requires a knowledge base, IRN-CDB, in which optimal configurations for the production network are stored. Optimal configurations are obtained for each system state $\phi[R_a]$. under the assumption that the system will remain in such state without further change. This allows formulating the problem as a flow (throughput) maximization problem with capacity and topology constraints, as follows

$$\max TP[IRN]$$

$$\text{s.t.:} \qquad \vartheta\left[r_n^P | \phi\left[r_n^P\right]\right] \geq \mu\left[r_n^P\right] \quad \forall r_n^P \in R_a$$

Flow balance constraints (dependent on IRN topology)

where $\vartheta\left[r_n^P | \phi\left[r_n^P\right]\right]$ is process r_n^P capacity given it is at state $\phi\left[r_n^P\right]$. In essence, this approach determines the bottleneck in each state $\phi[R_a]$ and sets all non-bottleneck process flow rates according to the bottleneck rate to cooperate in achieving maximum throughput. As for storages, target levels are set to low values, ideally zero. The results of 2^n combinations of IRN state (taking into account only processes) are stored in IRN-CDB for later use by IFCP.

IFCP decisions rely on prognostic information provided by ECDT, which needs to be tailored specifically for each application. ECDT provides production network-level situation awareness, as it integrates information from distributed resources into a single evaluation of possible future system evolution. In the case of production networks, agent a may collect data regarding process r_n^P failures, namely time of occurrence and duration of each event. Let $F[r]$ be a set of failure observations for resource r collected by agent a; then, the latter may estimate the distribution of time between failures for r_n^P using the information collected.

$$t_{b/w fail}\left[r_n^P\right] \sim f\left(F\left[r_n^P\right]\right)$$

Let $t_{last fail}\left[r_n^P\right]$ be the failure time for r_n^P in $F\left[r_n^P\right]$. Then, agent a can obtain the probability $p_{ECDT}\left[r_n^P\right]$ that a failure occurs within a prognostic horizon Δt_{ECDT}, based on $t_{b/w fail}\left[r_n^P\right]$.

$$p_{ECDT}\left[r_n^P\right] = t_{b/w fail}\left[r_n^P\right]\left(x \leq t + \Delta t_{ECDT} - t_{last fail}\left[r_n^P\right]\right)$$

The level of situation awareness of agent a w.r.t. status of its internal resources is influenced by the amount, precision, and detail level of the failure data collected, as well as by the accuracy of its predictive model $p_{ECDT}\left[r_n^P\right]$. Comparison of $p_{ECDT}\left[r_n^P\right]$ with pre-defined threshold values $p\left(\phi\left[r_n^P\right] \in \mathbb{D}\right)$ define the need to introduce modifications to the set of control logic and interaction policies in order to increase the disruption readiness of the network and provide the next expected-to-fail process a better condition to face the likely disruption.

Finally, teaming rules (Fig. 10.2) for disruption readiness (i.e., while IRN is operating normally but a disruption is likely within the ECDT horizon), response, and recovery are defined to enable collaboration among resources in the production network.

IRN teaming rules for disruption readiness

$$\exists\, r_n^P \in R_a \mid p_{ECDT}[r_n^P] \ge p(\phi[r_n^P] \in \mathbb{D})$$
$$\rightarrow \begin{bmatrix} \text{Increase } p_{NB}(r_{n^*}^P) \text{ by reducing } TL[r_m^S] \; \forall n^* \text{ upstream of } n \\ \text{Increase } p_{NS}(r_{n'}^P) \text{ by increasing } TL[r_m^S] \; \forall n' \text{ downstream of } n \end{bmatrix}$$

IRN teaming rules for disruption response

Disrupted storage $r_m^S \in R_a$

$$\text{If } \vartheta[r_m^S](t) > TL[r_m^S] \rightarrow \begin{bmatrix} \text{Reduce } \vartheta[r_m^S](t) \text{ by increasing } \mu[r_{n'}^P] \; \forall n' \text{ downstream of } n \\ OR \text{ (if former is not possible)} \\ \text{Reduce } \vartheta[r_m^S](t) \text{ by decreasing } \mu[r_{n^*}^P] \; \forall n^* \text{ upstream of } n \end{bmatrix}$$

$$\text{If } \vartheta[r_m^S](t) < TL[r_m^S] \rightarrow \begin{bmatrix} \text{Increase } \vartheta[r_m^S](t) \text{ by increasing } \mu[r_{n^*}^P] \; \forall n^* \text{ upstream of } n \\ OR \text{ (if former is not possible)} \\ \text{Increase } \vartheta[r_m^S](t) \text{ by decreasing } \mu[r_{n'}^P] \; \forall n' \text{ downstream of } n \end{bmatrix}$$

Disrupted process $r_n^P \in R_a$

$$\text{While } \exists\, \phi[r_n^P] \in \mathbb{D} \rightarrow \begin{bmatrix} \text{Set } TL[r_m^S] = \vartheta[r_m^S] \; \forall n^* \text{ upstream of } n \\ \text{Set } TL[r_m^S] = 0 \; \forall n' \text{ downstream of } n \end{bmatrix}$$

Fig. 10.2 Process and storage teaming to anticipate and respond to disruptions

10.3 Design of Experiments

Based on a real-world tissue paper converting line (Fig. 10.3), a simulation model was built ARENA to compare the performance of IFCP to other known control methods in terms of QoS, throughput variability, and WIP. The simulated line is composed of five different processes, arranged in series and in parallel, and followed by a storage resource. Resource capacity $\vartheta[r]$ depends on flow characteristics, equipment design, and maintenance conditions. Successors of agent a consume finished products from the most downstream storage resource in the production network, whereas predecessors P_a source raw materials (RMs) to the most upstream storage.

A list of the most relevant resource control logic and interaction policies are presented in Table 10.1. Column resources indicates resources involved (logic/policy modifies behavior of the resource in the first line), trigger condition(s), and action(s) implemented either by automation or manually by an operator.

The simulated production line is configured to process two types of products (flows), one per each output branch (i.e., bagger and bundler combination). Parameters of the tissue converting line resources used in all experimental configurations are presented in Table 10.2. All instances of each process type in the tissue production line are assumed to be identical, i.e., having the same parameters and transformation function $f_r^{I/O}$ (with the exception of the processing rate, which depends on the type of flow processed at baggers and bundlers).

Capacity values have been scaled to preserve privacy and confidentiality of company information; storage capacities are scaled w.r.t. logsaw storage capacity

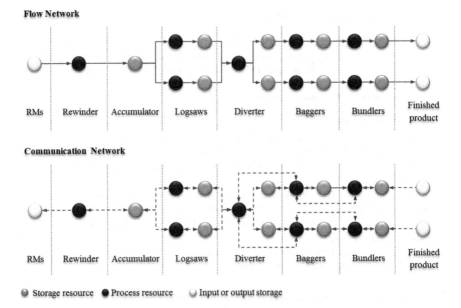

Flow Network

RMs   Rewinder   Accumulator   Logsaws   Diverter   Baggers   Bundlers   Finished product

Communication Network

RMs   Rewinder   Accumulator   Logsaws   Diverter   Baggers   Bundlers   Finished product

⬤ Storage resource ● Process resource ◡ Input or output storage

Fig. 10.3 Tissue converting line: Flow and communication networks

and process capacities w.r.t. logsaw capacity. Process failures were modeled using exponential distributions; mean time between failures (MTBF) and mean time to repair (MTTR), both in minutes, are indicated for each process in Table 10.2.

Production network performance is measured by QoS delivered to agents in S_a, throughput variability, and average WIP. Let $\mu[S_a](t)$ be the instantaneous flow rate demand of agents in S_a and $TP(t)$ the converting line throughput at time t; then, QoS is measured as follows

$$QoS_{a \to S_a} = \frac{\int_{t_{wu}}^{t_{run}} \min\{\mu[S_a](t), TP(t)\}dt}{\int_{t_{wu}}^{t_{run}} \mu[S_a](t)dt}$$

where t_{wu} and t_{run} are the simulation warm-up time and total run time, respectively. Similarly, WIP is measured as follows

$$E[WIP] = \sum_{r^S \in R_a} (t_{run} - t_{wu})^{-1} \int_{t_{wu}}^{t_{run}} \vartheta[r^S](t)dt$$

Throughput changes in discrete amounts over time period $[t_{run}; t_{wu}]$; hence, it is possible to obtain the probability distribution for throughput values over $[t_{run}; t_{wu}]$, namely $p(TP = x)$. Then, throughput variability can be measured by the standard deviation of TP, as follows

Table 10.1 Interaction policies

Control logic and interaction policies	Details			
	Type	Resources	Trigger condition(s)	Action(s)
Rewinder stop w/ full accumulator	A	Rewinder (P) Accumulator (S)	Accumulator is full	Stop rewinder
Rewinder re-start	A	Rewinder (P) Accumulator (S)	Accumulator below restart level after rewinder stop	Enable rewinder to opérate
Log production	A	Rewinder (P) Accumulator (S)	Rewinder not failed Accumulator not full	Produce 1 log
Rewinder speed control	A/M	Rewinder (P) Accumulator (S)	Accumulator level over/under target range	Adjust down/up rewinder speed
Accumulator unload	A	Accumulator (S) Logsaw (P)	Accumulator not empty Logsaw not full	Unload 1 log
Roll production	A	Logsaw (P) Logsaw storage (S)	Logsaw not failed Logsaw storage not full	Cut 1 roll
Logsaw speed control	A/M	Accumulator (S) Logsaw (P) Logsaw storage (S)	Accumulator level out of target range (over/under)	Adjust logsaw speed (up/down)
Diverter load/unload sequence	A	Logsaw storage (S) Diverter (P) Diverter storage (S)	Diverter storage not full Logsaw storage not empty Diverter not failed	Unload 1 diverter batch
Bagger/Diverter failure interaction	A	Bagger (P) Diverter (P)	Bagger failed	Disable diverter unload to failed bagger
Pack production	A	Bagger (P) Diverter storage (S) Bagger storage (S)	Bagger not failed Diverter storage not empty Bagger storage not full	Produce 1 pack
Bagger speed control	A/M	Bagger (P) Diverter storage (S) Bagger storage (S)	Diverter storage out of target range (over/under)	Adjust bagger speed (up/down)
Bagger/Bundler failure interaction	A	Bagger (P) Bundler (P)	Bundler failed	Disable bagger feed to failed bundler
Bundle production	A/M	Bundler (P) Bundler storage (S)	Bundler not failed Bundler storage not empty	Produce 1 bundle

A automated; *M* manual; *P* process; *S* storage

Table 10.2 Tissue converting line resource parameters

Process	Capacity	Failures		Storage	Capacity
		MTBF	MTTR		
Rewinder	2	60	10	Accumulator	90
Logsaw	1	1440	25	Logsaw storage	1
Diverter	3.5	–	–	Diverter storage	1
Bagger	1.3	240	20	Bagger storage	(5.5; 11.5)
Bundler	3	600	20	Bundler storage	12

$$SD[TP] = \sum_x (x - E[TP])^2 p(TP = x)$$

where

$$E[TP] = (t_{run} - t_{wu})^{-1} \int_{t_{wu}}^{t_{run}} TP(t)dt.$$

Using the abovementioned simulation model, five different flow control protocols were tested: Base-stock, CONWIP, DWIP, Kanban, and IFCP. In each case, the control parameters were adjusted heuristically in order to achieve the highest possible production network performance. A period t_{run}. of one week following a warm-up period t_{wu}. of one day was replicated 100 times to obtain statistically valid inferences on the indicators.

10.4   Case Study Results and Discussion

The results from the experiments described in the preceding section with $\mu[S_a](t) = 7.3$ cases/min are presented in Table 10.3 and Fig. 10.4. In the case of IFCP, the results correspond to parameters $p_{ECDT}[r^P] = 0.7 \, \forall r^P \in R_a$ and $\Delta t_{ECDT} = 0$.

From the simulation results, it can be inferred that the control policies can be clustered in three groups based on QoS, throughput variability, and WIP: (1) Base-stock and CONWIP, (2) DWIP and Kanban, and (3) IFCP. In order to validate this inference, two-sample unpooled t-tests assuming unequal variances were performed to compare the different control protocols based on QoS and $E[WIP]$, with a significance level of 1%. Results are summarized in Table 10.4; each cell shows the result of the hypothesis test comparing the protocol corresponding to the row with the protocol corresponding to the column, for the performance metric. Statistical significance indicates that the protocol corresponding to the cell's row outperforms the protocol corresponding to the column.

Table 10.3 Simulation results

Control Protocol	QoS	$SD[TP]$ (un./min)	$E[WIP]$ (un.)
Base-stock	99.390	0.0816	157.85
CONWIP	99.361	0.0941	164.24
DWIP	99.475	0.0766	130.33
Kanban	99.492	0.0782	124.75
IFCP	99.396	0.0436	92.36

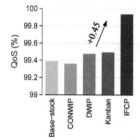 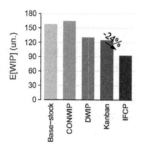 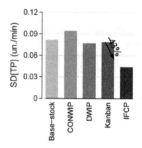

Fig. 10.4 Comparison of QoS, WIP, and throughput variability

Statistical comparison of QoS resulting from each control approach allows concluding that IFCP outperforms Base-stock, CONWIP, DWIP and Kanban with a significance level of 1%, yielding a 0.45% increase in QoS. Furthermore, hypothesis tests show it is not possible to conclude that there are statistically significant differences in QoS within the group of Kanban, DWIP, CONWIP, and Base-stock.

Analysis of results in terms of $E[WIP]$ show statistically significant differences among three clusters: IFCP (lowest $E[WIP]$ level), Kanban and DWIP (intermediate $E[WIP]$ level), and CONWIP and Base-stock (highest $E[WIP]$ level). A 24% reduction in $E[WIP]$ is obtained by using IFCP instead of either method in the intermediate $E[WIP]$ level cluster (Kanban or DWIP).

In this application, the objective of IFCP was threefold: improve (1) QoS, while simultaneously reducing (2) throughput variability and (3) WIP. From the simulation results, it can be inferred that IFCP reduces variability, measured as the standard deviation of throughput, $SD[TP]$. To statistically validate this conclusion, a one-sided F-test was performed to compare the standard deviation of different control protocols, with a significance level of 1%.

Results are summarized in Table 10.5; each cell shows the result of the hypothesis test comparing the protocol corresponding to the row with the protocol corresponding to the column. Statistical significance indicates that the protocol corresponding to the cell's row outperforms the protocol corresponding to the column, i.e., the variance of the latter is larger than that of the row's protocol. The statistical comparison shows that IFCP yields a 43% reduction in throughput variance, hence a lower variability, when compared to alternative control methods.

Table 10.4 QoS and WIP—Hypothesis test results at 0.99 confidence level (one sided)

Control Protocol	QoS				$E[WIP]$			
	Base-stock	CONWIP	DWIP	Kanban	Base-stock	CONWIP	DWIP	Kanban
IFCP	X	X	X	X	X	X	X	X
Kanban	–	–	–		X	X	–	
DWIP	–	–			X	X		
CONWIP	–				–			

X Statistically significant difference; – No statistically significant difference

Table 10.5 Throughput variability—Hypothesis test results at 0.99 confidence level (one sided)

Control Protocol	$SD[TP]$			
	Base-stock	CONWIP	DWIP	Kanban
IFCP	X	X	X	X
Kanban	–	–		
DWIP	–	–		–
CONWIP				
Base-stock		–		

X statistically significant difference (column variance > row variance); – No statistically significant difference

The above-discussed experimental results clearly indicate that production network control via IFCP outperforms traditional control methodologies in terms of QoS and throughput variability. Without additional or more reliable resources, IFCP is able to effectively team processes and storages to provide higher QoS with lower throughput variability. Furthermore, IFCP's anticipatory capabilities enable a more efficient use of available storage in order to provide higher protection against disruptions while simultaneously reducing WIP. Based on these observations, it can be concluded that FTT-based flow control is able to increase resilience to resource failures in IRNs. Moreover, achieving higher resilience without increasing structural costs i.e., investment in additional or more reliable resources, and reducing operational costs, i.e., tied-up capital in the form of WIP and its associated maintenance expenses, provides the base for increased long-term sustainability.

References

Dalal, M., Groel, B., & Prieditis, A. (2003). Real-time decision making using simulation. In *Proceedings of the 2003 Winter Simulation Conference* (pp. 1456–1464) New Orleans, LA, USA. doi:10.1109/WSC.2003.1261589

Hudson, S., McNamara, T., & Shaaban, S. (2014). Unbalanced lines: Where are we now? *International Journal of Production Research, 53*(6), 1895–1911. doi:10.1080/00207543.2014.965357

Ma, Y. H. K., & Koren, Y. (2004). Operation of manufacturing systems with work-in-process inventory and production control. *CIRP Annals-Manufacturing Technology, 53*(1), 361–365. doi:10.1016/S0007-8506(07)60717-3

Paternina-Arboleda, C. D., & Das, T. K. (2001). Intelligent dynamic control policies for serial production lines. *IIE Transactions, 33*(1), 65–77. doi:10.1080/07408170108936807

Reyes Levalle, R. (2015). Resilience by teaming in supply networks. Purdue University.

Reyes Levalle, R., Scavarda, M., & Nof, S. Y. (2013). Collaborative production line control: Minimisation of throughput variability and WIP. *International Journal of Production Research, 51*(23–24), 7289–7307. doi:10.1080/00207543.2013.778435

Yang, R. L., Subramaniam, V., & Gershwin, S. B. (2006). Setting real time WIP levels in production lines, Innovation in Manufacturing Systems and Technology (IMST). Singapore-MIT Alliance.

Chapter 11
Case Study B: Network Formation and Flow Control Protocols Applied to Physical Distribution Networks

Flow delivery problems in distribution networks have received significant attention from researchers, as these challenges are ubiquitous in modern physical supply networks. Initially, research focused mostly on vehicle routing problems (VRPs), a sub-set of flow delivery problems concerned with the selection of intermediaries and delivery paths to minimize the total cost of delivery. As QoS gained importance, approaches to the solution of VRPs began to incorporate pre-negotiated delivery lead times as constraints to ensure fulfillment of SLAs. However, in practice, dynamic network conditions oftentimes hinder static solutions if these do not incorporate enough slack to allow for unexpected delays in the delivery plan.

Small parcel distribution networks have grown significantly over the last years, mainly fueled by the exponential growth of e-commerce driven by Amazon and store pickup and home delivery options offered by large retailers such as Walmart, Target, and Best Buy, among others. Over time, increasing pressure from these systems' customers to obtain best-in-class QoS with shorter leadtimes and lower cost eroded the margin of static solutions, obtained through traditional VRP approaches, to cope with congestion and/or disruption in flow links and/or at intermediaries. Furthermore, some SN agents became increasingly unwilling to accept tardiness, even in the event of unforeseen events, often including penalties in SLAs for late deliveries. In this new scenario, resilience-enabling distribution protocols based on teaming (DNF/RP and DFCP), capable of adapting dynamically to changing distribution network conditions, outperform traditional approaches based on cost and/or leadtime minimization (Reyes Levalle 2015a, b).

© Springer International Publishing AG 2018
R. Reyes Levalle, *Resilience by Teaming in Supply Chains and Networks*,
Automation, Collaboration, & E-Services 5, DOI 10.1007/978-3-319-58323-5_11

11.1   Application of DNF/RP and DFCP to a Small Parcel Delivery Network

Parcel delivery systems involve three types of agents: distribution centers (DC), hubs or cross-docking facilities (H), and destinations (D). Distribution centers store the flow that needs to be delivered to destination agents, which may be individual customers, in large network representations, or, more commonly, local warehouses of last-mile delivery services. Hubs are intermediary agents that provide connectivity between DCs and several destinations. Following the definitions introduced Sect. 9, a DC is an origin agent i that supplies several destination agents j through flow links let $fl_{i \to j}$. Each flow link is operationalized by a series of interconnections, between i and each j, and a set of hubs $h \in H_{i \to j}$, for each connection.

In order to apply DNF/RP and DFCP to a parcel distribution network, a set of fuzzy scoring rules and fuzzy input variables are required. Figure 11.1 (left) presents the membership functions for three fuzzy variables, namely LOW, MEDIUM, and HIGH, which can be parameterized by selecting $MF_x, x = 1, \ldots, 4$. Then, a set of MF_x values must be selected for each input metric $Slack\left[\delta_{i \to j}\right]$, $Cost\left[\delta_{i \to j}\right]$, and $Flex\left[\delta_{i \to j}\right]$, in order to convert crisp values into their fuzzy counterparts, with their corresponding degree of membership (DOM).

Using fuzzy measures obtained from real-time data for $Slack\left[\delta_{i \to j}\right]$, $Cost\left[\delta_{i \to j}\right]$, and $Flex\left[\delta_{i \to j}\right]$ values, a set of fuzzy rules is applied to produce a score for path $\delta_{i \to j}$. Fuzzy rules (Table 11.1) are designed to evaluate paths based on the degree of fulfillment (DOF) of the condition stated by the IF...THEN rule. DOF is obtained by computing the product of each criterion's DOM. For instance, if "$Slack\left[\delta_{i \to j}\right]$ isLOW" with DOM 0.5, "$Cost\left[\delta_{i \to j}\right]$ isLOW" with DOM 0.8, and "$Flex\left[\delta_{i \to j}\right]$ isLOW" with DOM 0.2, the degree of fulfilment of the condition in Rule #1 is 0.08.

Depending on the implication operator chosen, the DOF of each condition will affect the shape of the fuzzy output score differently. Fuzzy output scores are shown in Fig. 11.1 (right). When all rules are processed for a given path $\delta_{i \to j}$, rules' output

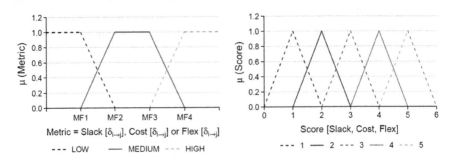

Fig. 11.1 Fuzzy input variables for $Slack\left[\delta_{i \to j}\right]$, $Cost\left[\delta_{i \to j}\right]$, and $Flex\left[\delta_{i \to j}\right]$, and fuzzy output $Score[Slack, Cost, Flex]$

Table 11.1 Fuzzy rules for DNF/RP and DFCP

Rule	Description
1	IF $Slack[\delta_{i \to j}]$ is **LOW** AND $Cost[\delta_{i \to j}]$ is **LOW** AND $Flex[\delta_{i \to j}]$ is **LOW** THEN *Score* is 1
2	IF $Slack[\delta_{i \to j}]$ is **LOW** AND $Cost[\delta_{i \to j}]$ is **LOW** AND $Flex[\delta_{i \to j}]$ is **MED** THEN *Score* is 1
3	IF $Slack[\delta_{i \to j}]$ is **LOW** AND $Cost[\delta_{i \to j}]$ is **LOW** AND $Flex[\delta_{i \to j}]$ is **HIGH** THEN *Score* is 2
4	IF $Slack[\delta_{i \to j}]$ is **LOW** AND $Cost[\delta_{i \to j}]$ is **MED** AND $Flex[\delta_{i \to j}]$ is **LOW** THEN *Score* is 1
5	IF $Slack[\delta_{i \to j}]$ is **LOW** AND $Cost[\delta_{i \to j}]$ is **MED** AND $Flex[\delta_{i \to j}]$ is **MED** THEN *Score* is 1
6	IF $Slack[\delta_{i \to j}]$ is **LOW** AND $Cost[\delta_{i \to j}]$ is **MED** AND $Flex[\delta_{i \to j}]$ is **HIGH** THEN *Score* is 2
7	IF $Slack[\delta_{i \to j}]$ is **LOW** AND $Cost[\delta_{i \to j}]$ is **HIGH** AND $Flex[\delta_{i \to j}]$ is **LOW** THEN *Score* is 1
8	IF $Slack[\delta_{i \to j}]$ is **LOW** AND $Cost[\delta_{i \to j}]$ is **HIGH** AND $Flex[\delta_{i \to j}]$ is **MED** THEN *Score* is 1
9	IF $Slack[\delta_{i \to j}]$ is **LOW** AND $Cost[\delta_{i \to j}]$ is **HIGH** AND $Flex[\delta_{i \to j}]$ is **HIGH** THEN *Score* is 1
10	IF $Slack[\delta_{i \to j}]$ is **MED** AND $Cost[\delta_{i \to j}]$ is **LOW** AND $Flex[\delta_{i \to j}]$ is **LOW** THEN *Score* is 3
11	IF $Slack[\delta_{i \to j}]$ is **MED** AND $Cost[\delta_{i \to j}]$ is **LOW** AND $Flex[\delta_{i \to j}]$ is **MED** THEN *Score* is 4
12	IF $Slack[\delta_{i \to j}]$ is **MED** AND $Cost[\delta_{i \to j}]$ is **LOW** AND $Flex[\delta_{i \to j}]$ is **HIGH** THEN *Score* is 4
13	IF $Slack[\delta_{i \to j}]$ is **MED** AND $Cost[\delta_{i \to j}]$ is **MED** AND $Flex[\delta_{i \to j}]$ is **LOW** THEN *Score* is 2
14	IF $Slack[\delta_{i \to j}]$ is **MED** AND $Cost[\delta_{i \to j}]$ is **MED** AND $Flex[\delta_{i \to j}]$ is **MED** THEN *Score* is 3
15	IF $Slack[\delta_{i \to j}]$ is **MED** AND $Cost[\delta_{i \to j}]$ is **MED** AND $Flex[\delta_{i \to j}]$ is **HIGH** THEN *Score* is 3
16	IF $Slack[\delta_{i \to j}]$ is **MED** AND $Cost[\delta_{i \to j}]$ is **HIGH** AND $Flex[\delta_{i \to j}]$ is **LOW** THEN *Score* is 1
17	IF $Slack[\delta_{i \to j}]$ is **MED** AND $Cost[\delta_{i \to j}]$ is **HIGH** AND $Flex[\delta_{i \to j}]$ is **MED** THEN *Score* is 2
18	IF $Slack[\delta_{i \to j}]$ is **MED** AND $Cost[\delta_{i \to j}]$ is **HIGH** AND $Flex[\delta_{i \to j}]$ is **HIGH** THEN *Score* is 2
19	IF $Slack[\delta_{i \to j}]$ is **HIGH** AND $Cost[\delta_{i \to j}]$ is **LOW** AND $Flex[\delta_{i \to j}]$ is **LOW** THEN *Score* is 5
20	IF $Slack[\delta_{i \to j}]$ is **HIGH** AND $Cost[\delta_{i \to j}]$ is **LOW** AND $Flex[\delta_{i \to j}]$ is **MED** THEN *Score* is 5
21	IF $Slack[\delta_{i \to j}]$ is **HIGH** AND $Cost[\delta_{i \to j}]$ is **LOW** AND $Flex[\delta_{i \to j}]$ is **HIGH** THEN *Score* is 5

(continued)

Table 11.1 (continued)

Rule	Description
22	IF $Slack\left[\delta_{i\rightarrow j}\right]$ is **HIGH** AND $Cost\left[\delta_{i\rightarrow j}\right]$ is **MED** AND $Flex\left[\delta_{i\rightarrow j}\right]$ is **LOW** THEN $Score$ is 4
23	IF $Slack\left[\delta_{i\rightarrow j}\right]$ is **HIGH** AND $Cost\left[\delta_{i\rightarrow j}\right]$ is **MED** AND $Flex\left[\delta_{i\rightarrow j}\right]$ is **MED** THEN $Score$ is 4
24	IF $Slack\left[\delta_{i\rightarrow j}\right]$ is **HIGH** AND $Cost\left[\delta_{i\rightarrow j}\right]$ is **MED** AND $Flex\left[\delta_{i\rightarrow j}\right]$ is **HIGH** THEN $Score$ is 4
25	IF $Slack\left[\delta_{i\rightarrow j}\right]$ is **HIGH** AND $Cost\left[\delta_{i\rightarrow j}\right]$ is **HIGH** AND $Flex\left[\delta_{i\rightarrow j}\right]$ is **LOW** THEN $Score$ is 3
26	IF $Slack\left[\delta_{i\rightarrow j}\right]$ is **HIGH** AND $Cost\left[\delta_{i\rightarrow j}\right]$ is **HIGH** AND $Flex\left[\delta_{i\rightarrow j}\right]$ is **MED** THEN $Score$ is 3
27	IF $Slack\left[\delta_{i\rightarrow j}\right]$ is **HIGH** AND $Cost\left[\delta_{i\rightarrow j}\right]$ is **HIGH** AND $Flex\left[\delta_{i\rightarrow j}\right]$ is **HIGH** THEN $Score$ is 3

scores are de-fuzzified using the center of sums method to produce a crisp score, $Score[\delta_{i\rightarrow j}]$. The center of sums is obtained by computing the weighted average of the centroids of each output score function, using the function's area as weight.

11.2 Design of Experiments

DNF/RP and DFCP were tested using a distribution network modeled in ARENA. The network consists of one DC located in Indianapolis, three hubs located in Lafayette (H1), Kokomo (H2), and Marion (H3), and six destinations located in Monticello (D1), Frankfort (D2), Logansport (D3), Peru (D4), Wabash (D5), and Fort Wayne (D6). Distances of flow links $fl_{a\rightarrow b}$, $a \in \{DC,H1,H2,H3\}$ and $b \in \{H1, H2, H3, D1, D2, D3, D4, D5, D6\}$, and transportation cost $\omega_{a\rightarrow b}$ between different nodes in the distribution network are shown in Table 11.2; a dash "-" indicates that no connection between nodes is permitted.

When an order arrives at a hub, it must undergo processing before it continues towards its final destination. Processing time at a hub, depends on the hub's processing rate $\mu[h]$ as well as the queue of orders pending to be processed. Hubs are subject to random disruptions during which they are unable to process orders; these must wait until the disruption is resolved to be routed to their final destination. Probability distributions for disruption frequency and duration distribution, and hub processing rate are presented in Table 11.3. Hub processing costs ω_h are assumed fixed and equal to 3, 4, and 3 for hubs H1, H2, and H3, respectively.

Maximum travel speed in flow link $fl_{a\rightarrow b}$, $a \in \{i\} \cup H_{i\rightarrow j}$ and $b \in \{j\} \cup H_{i\rightarrow j}$ depends on the type of road that connects the nodes; 60 mph is assumed for interstate highways and 50 mph for other road types (Table 11.4). In order to model route congestion, maximum speeds are reduced by a random factor to produce instantaneous speeds; 70% of the time routes are congestion-free and 30% have a congestion between 0 and 83%, i.e., instantaneous speed can be as low as 17% of the maximum speed.

Orders from each destination occur at random intervals; each time an order is placed, the destination agent defines a requested leadtime for delivery, RLT. Table 11.5 presents the probability distributions for order frequency and RLT, for each destination agent.

Table 11.2 Distances and route use cost

From	To								
	H1	H2	H3	D1	D2	D3	D4	D5	D6
DC	63;13	51;10	81;11	–	–	–	–	–	–
H1	–	40;3	–	30;3	25;3	38;5	54;7	–	–
H2	40;3	–	30;6	–	34;6	24;3	21;2	38;1	–
H3	–	30;6	–	–	–	46;6	30;4	19;2	55;9

Key: distance (miles); cost $\omega_{a\rightarrow b}$($/use)

Table 11.3 Hub disruption frequency and duration, processing time, and cost

	Disruptions		Processing time	Processing cost
Hub	Frequency (hrs.)	Duration (min.)	(min./order)	($/use)
H1	Expo(9)	Unif(45; 75)	Unif(10; 15)	3
H2	Expo(12)	Tria(45; 60; 90)	Unif(10; 20)	4
H3	Expo(18)	Tria(25; 35; 45)	Unif(10; 15)	3

Expo(mean), Unif(minimum; maximum), Tria(minimum; mode; maximum)

Table 11.4 Maximum travel speeds for each flow link, in mph

	To								
From	H1	H2	H3	D1	D2	D3	D4	D5	D6
DC	60	50	60	–	–	–	–	–	–
H1	–	50	–	50	60	50	50	–	–
H2	50	–	50	–	50	50	50	50	–
H3	–	50	–	–	–	50	50	50	60

Table 11.5 Demand parameters for each destination

Destination	Order frequency (min.)	RLT (hrs.)
D1	Expo(86)	Unif(3.8; 4.2)
D2	Expo(43)	Tria(4.5; 5; 5.5)
D3	Expo(43)	Unif(4; 5)
D4	Expo(43)	Unif(4.5; 5)
D5	Expo(43)	Tria(4; 4.25; 5)
D6	Expo(86)	6

Expo(mean), Unif(minimum; maximum), Tria(minimum; mode; maximum)

The parcel distribution network and control protocols DNF/RP and DFCP were coded in ARENA (user interface is presented in Fig. 11.2). Real-time simulation data is captured from the model to feed protocols and select paths to anticipate and avoid congestion and disruptions. Data gathered from the system provides each agent with situation awareness; it is assumed that all agents can access data from all components of the distribution network, upon request. Route flexibility is assessed considering a distance relevance weight κ of 0.2. In order for DNF/RP and DFCP to dynamically re-evaluate paths under changing network conditions, ρ_{DC} must be under 50% of the initial $Slack[\delta_{i \to j}]$.

Parameters of fuzzy input variables for $Slack[\delta_{i \to j}]$, $Cost[\delta_{i \to j}]$, and $Flex[\delta_{i \to j}]$ are presented in Table 11.6. Using VBA modules, Mamdani-min and Larsen Product implication operators were used to compute output scores for fuzzy rules based on the DOF.

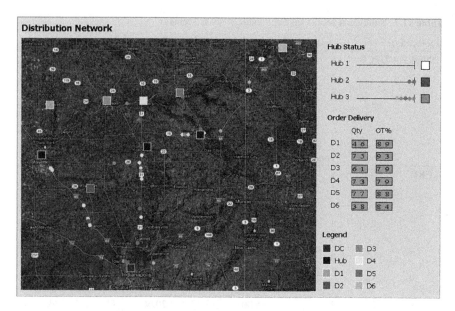

Fig. 11.2 ARENA simulation model (user interface) *Each circle on an interstate/highway shows its destination; dashboard indicates H2 is disrupted while H1 is unused and H3 is operational

Table 11.6 Fuzzy membership functions' parameters

	Parameters			
Metric	MF1	MF2	MF3	MF4
Slack	−1	2	4	5
Cost	13	18	20	15
Flexibility	0	0.5	1.5	2.5

With the purpose of benchmarking performance of DNF/RP and DFCP, two alternative simulation models were coded using (1) shortest time and (2) lowest cost routing protocols, STRP and LCRP, respectively.

11.3   Case Study Results and Discussion

Using the simulation models for DNF/RP + DFCP (with Mamdani-min operator), STRP, and LCRP, 15 replications of 8 days of operation were run and data was collected for the last 7 days of simulation for steady state assessment. Summary results for requested leadtime (RLT), delivery leadtime (DLT), total cost (TC), percentage of on-time deliveries (OT), and percentage of deliveries that required re-routing (RR) are presented in Table 11.7.

Table 11.7 Simulation results (average values)—base scenario

Protocol(s)	Destination	RLT	DLT	TC	OT (%)	RR (%)
DNF/RP + DFCP	1	4.00	3.19	19.00	77	0
	2	5.00	3.18	19.85	90	23
	3	4.50	3.44	20.89	83	34
	4	4.75	3.31	19.63	87	32
	5	4.25	3.20	16.93	86	22
	6	6.00	4.28	23.00	86	0
	Total (All)	4.70	3.37	19.65	86	22
LCRP	1	4.00	4.33	19.00	50	0
	2	5.00	4.17	19.00	72	0
	3	4.51	10.15	18.60	14	40
	4	4.75	9.42	18.78	31	40
	5	4.25	9.63	17.38	30	40
	6	6.00	4.12	23.00	85	0
	Total (All)	4.70	7.52	18.95	43	24
STRP	1	4.00	3.11	19.00	79	0
	2	5.00	3.37	20.66	87	33
	3	4.50	3.43	21.78	82	39
	4	4.75	3.39	21.26	86	40
	5	4.25	3.17	18.11	85	33
	6	6.00	4.26	23.00	85	0
	Total (All)	4.70	3.41	20.56	85	29

Performance of each control protocol was evaluated based on the percentage of on-time deliveries, total cost, and number of times re-routing was required to overcome disruptions and congestion. Aggregate results for each protocol (Fig. 11.3) show that DNF/RP and DFCP are able to provide marginally higher OT that STRP (86 vs. 85%) with a 4.5% lower total cost (20.56 vs. 19.65, validated through a hypothesis test with 99% confidence level). As for LCRP, it is able to provide the lowest routing cost (18.95); however, OT drops to 43%. A closer analysis of the performance difference between DNF/RP + DFCP and STRP shows that OT is similar for each destination; however, the number of deliveries requiring re-routing is 7% lower when teaming protocols are used, resulting in a lower overall cost.

Statistical comparison of average performance in OT, RR, and TC is show in Table 11.8. Hypothesis tests comparing the control protocol shown on each row vs. the protocol corresponding to each column were performed with a significance level of 1%. One-sided comparisons allow assessing whether row protocols show higher OT, lower RR, and lower TC than column protocols. Results show that DNF/RP + DFCP outperform LCRP in terms of on-time delivery, and STRP in relation to percentage of deliveries requiring re-routing and total cost.

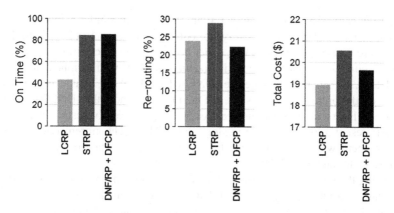

Fig. 11.3 Aggregate results for OT, RR, and TC

Table 11.8 Control protocol performance—hypothesis test results at 0.99 confidence level (one sided)

Control protocol	OT		RR			TC
	LCRP	STRP	LCRP	STRP	STRP	DNF/RP + DFCP
LCRP			X	X	X	
STRP	X					
DNF/RP + DFCP	X	–	–	X	X	

X statistically significant difference, – no statistically significant difference

To evaluate the effect of congestion level on protocol performance, average order generation rate for each destination was modified at 10% increments, from 70 to 120%, versus. base scenario. Simulation experiments were repeated using the DNF/RP + DFCP (with Mamdani-min operator), STRP, and LCRP models, running 15 replications of 8 days of operation and collecting data from the last 7 days of simulation for steady state assessment. Based on the results, OT, RR, and TC response curves (Fig. 11.4) were built for DNF/RP + DFCP and STRP, in order to assess how each protocol is able to cope with various levels of congestion at hubs.

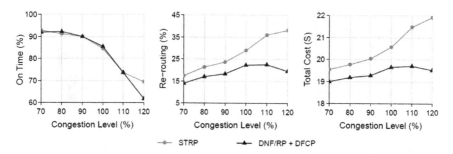

Fig. 11.4 OT, RR, and TC congestion response curves for DNF/RP + DFCP and STRP

Results show that DNF/RP + DFCP and STRP have similar responses to congestion in terms of on-time delivery, except for congestion levels beyond 110%. Protocol response diverges because, under high congestion scenarios (i.e., when all hubs are congested), leadtime slack is fuzzified as LOW with membership function equal to 1, for all paths. Therefore, DNF/RP + DFCP are unable to adequately decide on the optimal path. Results in Fig. 11.4 also show that DNF/RP + DFCP are able to limit total cost and the need for re-routing when congestion increases (or at least present a lower increase in slope than STRP). The aforementioned results are consistent with those observed by other approaches that consider re-routing, such as Wang and Shi (2009).

Finally, the results for DNF/RP + DFCP with Mamdani-min and Larsen product implication operators were compared under the base scenario, to assess the impact of condition DOF scoring on protocol behavior. The Larsen product operator has a deeper effect on the area under the output fuzzy variable than Mamdani-min; therefore, it enhances the weight of rules with high DOF over those with low DOF when using the center-of-sums to de-fuzzify output scores. Results (Fig. 11.5) indicate there is no statistically significant difference (at 99% confidence level) between protocol performances when using different implication operators.

The above-discussed results confirm that FTT-based design and control of flow distribution can increase resilience of delivery operations while simultaneously ensuring their long-term sustainability. When compared to STRP, FTT-based path design through DNF/RP is able to incorporate the teaming ability of chosen intermediaries as a decision factor through $Flex\left[\delta_{i \to j}\right]$. Hence, when leadtime slack is low, paths with higher teaming capability are able to better cope with unforeseen congestion and disruptions, increasing the number of on-time deliveries versus STRP. To balance any increase in cost paid for higher flexibility when needed, FTT-based design protocols select lower performance low-cost paths when available leadtime slack is higher. This results in lower RR percentages as no paths are altered by DFCP when sufficient slack is available, even after using part of the available leadtime slack to overcome congestion and disruptions. Using lower cost options, whenever possible, reduces operational cost versus STRP, making the combination of FTT-based protocol DNF/RP and DFCP more sustainable in the long-term.

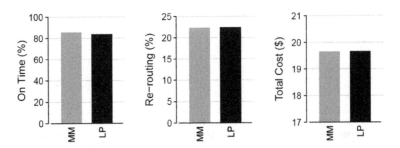

Fig. 11.5 OT, RR, and TC for DNF/RP + DFCP with Mamdani-Min (MM) and Larsen Product (LP) operators

Results indicate that, for a given FR_a, there exists a critical congestion level over which current FR_a fails to effectively leverage network teaming capabilities, resulting in a significant drop in performance versus STRP. Hence, to ensure the resilience and sustainability benefits of FTT-based design and operation protocols for flow distribution are captured throughout a wide range of congestion scenarios, FR_a must be re-calibrated accordingly.

References

Reyes Levalle, R., & Nof, S. Y. (2015a). Resilience by teaming in supply network formation and re-configuration. *International Journal of Production Economics, 160,* 80–93. doi:10.1016/j.ijpe.2014.09.036

Reyes Levalle, R., & Nof, S. Y. (2015b). A resilience by teaming framework for collaborative supply networks. *Computers & Industrial Engineering, 90,* 67–85. doi:10.1016/j.cie.2015.08.017

Wang, Z. & Shi, J. (2009). A model for urban distribution system under disruptions of vehicle travel time delay. In *Proceedings of the 2nd international conference on intelligent computing technology and automation* (pp. 433–436). Dalian, China. doi:10.1109/ICICTA.2009.570

Chapter 12
Case Study C: Beyond Agent-Level Benefits—The Effect of Resilience by Teaming on Network-Level Resilience

Agent-level resilience mechanisms received significant attention over the last years. A wide variety of approaches have been developed based on a trade-off between increased resilience and higher redundancy and use of excess resources. Despite the capacity of some approaches to increase agents' resilience, their relative benefits have seldom been analyzed from a global perspective. Research in complex systems has successfully linked network formation/re-configuration phenomena, driven by agent-level association/dissociation decisions, and the impact of topology on network performance and survivability (Barabási and Albert 1999; Albert et al. 2000; Albert and Barabási 2002; Tangmunarunkit et al. 2002; Thadakamalla et al. 2004; Brede and de Vries 2009). Nevertheless, decision mechanisms analyzed mostly rely on simple probabilistic association/dissociation rules and fail to account for more complex decision criteria involved in supply network formation and re-configuration processes.

Regardless of the magnitude of the benefits obtained from any agent-level strategies, it is not possible to extrapolate these benefits to the performance of the entire SN. Furthermore, it is necessary to understand if local benefits are obtained at the expense of performance losses in other SN agents, and whether these approaches create vulnerabilities that hinder SNs ability to be resilient. This chapter analyzes and discusses the benefits of RBT-based decisions relative to traditional SN formation and re-configuration mechanisms based on the results of Reyes Levalle (2015) and Reyes Levalle and Nof (2015a). Results show how local teaming-driven decisions effectively modify SN topology and its ability to overcome disruptions with minimal impact on the SN objectives.

© Springer International Publishing AG 2018
R. Reyes Levalle, *Resilience by Teaming in Supply Chains and Networks*,
Automation, Collaboration, & E-Services 5, DOI 10.1007/978-3-319-58323-5_12

12.1 Network Formation and Re-Configuration Dynamics

Every SN agent seeks to maximize its own benefits obtained from the interaction with other agents. As defined in Sect. 2.2.2, each agent $a \in A$ must decide which flow and communication links to maintain with other agents $i, j \in A$; a process that defines two sets of agents associated with a, namely its predecessors, P_a, and successors, S_a.

$$P_a = \{i | \exists fl_{i \to a} \in FL \text{ and } \exists cl_{a \to i} \in CL, i, a \in A\}$$

$$S_a = \{j | \exists fl_{a \to j} \in FL \text{ and } \exists cl_{j \to a} \in CL, a, j \in A\}$$

Interaction conditions between agents are defined in a SLA. Let $Q_{i \to j}$ be the flow quantity i agrees to send to j under a predefined SLA; then, the QoS agent i provides to agent j, $QoS_{i \to j}$, can be measured by the fraction of flow $Q_{i \to j}$ that meets the conditions of the SLA. Although in many real-world SNs multiple flows can co-exist within a given SN agent, for simplicity and without loss of generality, it is assumed in this case study that each agent can process only one type of flow. This implies that each agent must receive sufficient input flow of a given type (from one or several predecessors), transform it into output flow of a (possibly) different type and send it to its successors. Therefore, the supply network can be multi-flow although each SN agent is constrained to process only one type of flow.

12.1.1 Selection of Predecessors

Real-world SNs may contain agents with a wide range of decision-making profiles. Nonetheless, at a basic level, three traditional strategies for predecessor selection can be identified in literature:

(1) SP—minCost Select a single predecessor to minimize cost
(2) SP—maxQoS Select a single predecessor to maximize the QoS received
(3) DP Select a primary predecessor to minimize cost (and allocate at least 50% of the sourcing volume) and a secondary predecessor to increase overall QoS received

Additionally, an agent may use RBT (Reyes Levalle and Nof 2015b), a teaming-based approach to selection and management of predecessors.

(4) *RBT* Select several low cost suppliers and team them to yield a combined high QoS.

Choice of predecessors $i \in P_a$ and allocation of flow quantities $Q_{i \rightarrow a}$ will determine agent a's sourcing cost, defined as

$$SC_a(P_a, Q_{i \rightarrow a}) = \sum_{i \in P_a} C_{i \rightarrow a} Q_{i \rightarrow a}$$

where $C_{i \rightarrow a}$ is a measure of the resources a needs to consume in order to receive a unit of flow from i.

Furthermore, protocols used by agent a to control internal resources and flow distribution affect its capacity to provide successors with higher QoS than that received from its predecessors $i \in P_a$. The extent to which a is capable of overcoming low QoS from a predecessor is characterized by parameter β_a, a measure of its resilience. For each predecessor i sending flow to agent a the effective QoS is $QoS_{i \rightarrow a}^{\frac{1}{\beta_a}}$. Values of $\beta_a > 1$ imply that the agent is capable of increasing the QoS received from a predecessor whereas $\beta_a < 1$ models the inability of the agent to overcome lower QoS. In the case of RBT agents, parameter β_a is affected by the number of predecessors; increasing the size of P_a leads to higher values of β_a.

Maximum achievable QoS by agent a depends on the strategy used to select predecessors. Table 12.1 presents the mathematical expression of QoS_a for each predecessor selection strategy. In all cases, QoS_a is independent of total input flow $\sum Q_{i \rightarrow a}$; however, for DP and RBT, it does depend on how total flow is allocated among predecessors. Also, should multi-sourcing agents select a single predecessor, their QoS_a matches the expression for single-sourcing agents.

$$\text{In DP:} \quad M \equiv S \Rightarrow QoS_{M \rightarrow a} = QoS_{S \rightarrow a} \Rightarrow QoS_a = QoS_{M \rightarrow a}^{\frac{1}{\beta_a}}$$

$$\text{In RBT:} \quad n = 1 \Rightarrow \beta_a(1) = \beta_a \Rightarrow QoS_a = QoS_{i \rightarrow a}^{\frac{1}{\beta_a}}$$

Agent benefit maximization depends on asynchronous decisions made to form and/or re-configure connections. Decision-making complexity is further increased

Table 12.1 Predecessor selection strategies, QoS limit, and bid selection rules

Strategy	P_a	QoS limit, QoS_a	Winning bid selection rule/s
SP— minCost	$i*$	$QoS_{i* \rightarrow a}^{\frac{1}{\beta_i}}$	$i* \leftarrow$ Select bid w/*min* $C_{i \rightarrow a}$
SP— maxQoS	$i*$	$QoS_{i* \rightarrow a}^{\frac{1}{\beta_a}}$	$i* \leftarrow$ Select bid w/*max* $QoS_{i \rightarrow a}$
DP	M, S	$\dfrac{Q_{M \rightarrow a} QoS_{M \rightarrow a}^{\frac{1}{\beta_a}} + Q_{S \rightarrow a} QoS^{\frac{1}{\beta_a}}}{Q_{M \rightarrow a} + Q_{S \rightarrow a}}$ where $Q_{S \rightarrow a} = (1 - QoS_{M \rightarrow a}) Q_{M \rightarrow a}$	$M \leftarrow$ Select bid w/*min* $C_{i \rightarrow a}$ $S \leftarrow$ Select bid w/*max* $QoS_{i \rightarrow a}$
RBT	$1, \ldots, n$	$\displaystyle\sum_{i=1}^{n} Q_{i \rightarrow a} QoS_{i \rightarrow a}^{\frac{1}{\beta_a(n)}} / \sum_{i=1}^{n} Q_{i \rightarrow a}$	Sort bids (decreasing) according to $\frac{QoS_{i \rightarrow a}}{C_{i \rightarrow a}}$ and select first n

by incomplete information, due to interdependencies among the distributed decisions of all SN agents, and randomness emerging from speculation. Therefore, no
centralized one-time optimization model can accurately describe the dynamics of
real SNs. Instead, bidding mechanisms, such as those presented in Ertogral and Wu
(2000) and Lee and Kumara (2007), are more suitable to describe these asynchronous processes and reach acceptable, or even near-optimal, solutions. Then, to
select predecessors, each agent initiates a first-price sealed-bid auction process.
Under this setting, bidding agents will not know their competitors' bids for a given
sourcing request. The winning bid is determined based on the auctioneer's strategy
towards predecessor selection, indicated in Table 12.1.

12.1.2 Selection of Successors

Successor and predecessor selection are asynchronous interdependent processes.
Link formation with a successor j requires agent a to submit a winning bid in an
auction process initiated by j. In order to maximize its benefits, i.e., rewards minus
sourcing cost, an agent participates in all possible sourcing auctions in an attempt to
maximize reward collection. However, agent's ability to submit a bid that results
attractive to an auctioneer is constrained by their sourcing cost and QoS limit at the
time the bid is sent (Fig. 12.1).

Each bid $b_{a \rightarrow j}$ agent a submits to an auction initiated by agent j contains two
parameters: $QoS_{a \rightarrow j}$ and unit flow cost $C_{a \rightarrow j}$. QoS offers made by agent a must be
bounded by QoS_a. Therefore,

$$QoS_{a \rightarrow j} \in [0, QoS_a] \forall j \in S_a$$

Unit flow cost $C_{a \rightarrow j}$ depends on the average sourcing cost SC_a of agent a,
$QoS_{a \rightarrow j}$ offered to j, flow volume $Q_{a \rightarrow j}$ requested by j, and a parameter π_a that

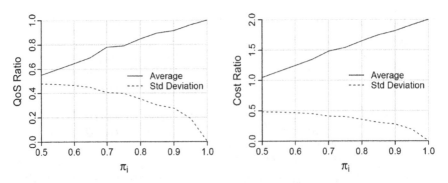

Fig. 12.1 QoS ratio and Cost ratio versus agent speculative behavior

defines a's speculative behavior regarding what its successors are willing to accept based on alternative sources available. Formally,

$$C_{a \to j} = f\left(\overline{SC_a}, QoS_{a \to j}, Q_{a \to j}, \pi_a\right)$$

where

$$\overline{SC_a} = \frac{SC_a}{\sum_{i \in P_a} Q_{i \to a}}$$

Every time a bid is accepted and a link $fl_{a \to j}$ is formed, agent a increases its total reward, defined as

$$RW_a\left(C_{a \to j}\right) = \sum_{j \in S_a} C_{a \to j} \, Q_{a \to j}$$

Link formation is immediately followed by a balance of input and output flow, either by readjusting input flows $Q_{i \to a}$ or modifying the predecessor set P_a.

$$\sum_{i \in P_a} Q_{i \to a} = \sum_{j \in S_a} Q_{a \to j}$$

12.1.3 Modeling Supply Network Formation and Re-Configuration Dynamics

In order to model association/dissociation dynamics leading to SN formation and re-configuration, each agent applies two protocols (Table 12.2) driven by a first-price sealed-bid auction mechanism:

(1) Network formation protocol (NFP): Creation of links to agents i which are neither included in P_a nor in set D of agents currently connected to a down-stream path that includes a
(2) Network re-configuration protocol (NRP): Re-allocation of flow among agents in P_a and/or elimination of links with agents $i \in P_a$.

Link creation/elimination and flow allocation procedures are presented in Fig. 12.2. In link formation/elimination, agent a calls for bids from agents $i \in B$. The latter must define their bids $b_{i \to a} = \{C_{i \to a}, QoS_{i \to a}\}$ and send it to agent a, which will select the winning bid/s and create/eliminate links accordingly. Bid quality of service $QoS_{i \to a}$ and unit flow cost $C_{i \to a}$ are defined as follows:

Table 12.2 Network formation protocol and network re-configuration protocol

Network formation protocol (NFP)	Network re-configuration protocol (NRP)
Initialize: $B, D = \emptyset$ For $j \in A, j \neq a$: If $\exists$ path starting at a and ending in j: $D \leftarrow D \cup \{j\}$ $B \leftarrow A \cap D^c \cap P_a^c$ Call Link Creation/Elimination procedure Call *Flow Allocation* procedure	Initialize: $B = \emptyset$ $B \leftarrow P_a$ Call Link Creation/Elimination procedure Call *Flow Allocation* procedure END
END	

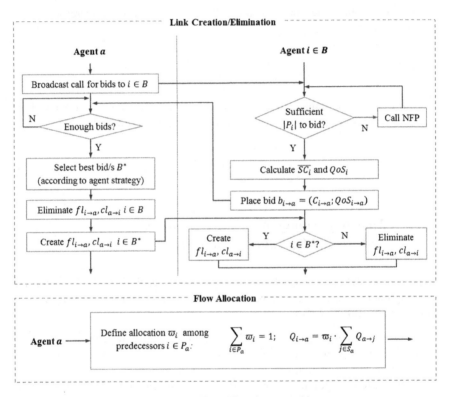

Fig. 12.2 Link Creation/Elimination and Flow Allocation procedures

$$QoS_{i \to a} = (1 - \Gamma)\eta_i \quad \Gamma = \begin{bmatrix} 0 & p = \pi_i \\ Unif(0, 0.1) & p = 1 - \pi_i \end{bmatrix}$$

$$C_{i \to a} = \overline{SC_i}\left(\pi_i + \frac{QoS_{i \to a}}{QoS_i}\right)$$

Parameter π_i models agent i's speculative behavior, and affects $C_{i \to a}$ and $QoS_{i \to a}$. In general, larger values of π_i, i.e., modeling lower speculation, increase

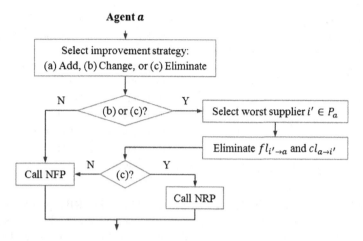

Fig. 12.3 SN agent total benefit improvement protocol

average $C_{i \to a}$ and $QoS_{i \to a}$, and reduce their standard deviation. Figure 12.1 shows the relation between π_i and ratios $QoS_{i \to a}/QoS_i$ (QoS ratio) and $C_{i \to a}/\overline{SC}_i$ (Cost ratio), from a sample of 100 observations.

During formation and re-configuration, changes to the set of predecessors P_a are constrained by current SLAs with successors; no modification to P_a should reduce QoS_a beyond the highest QoS among agent a's successors.

$$P_a \to P_a^* \Leftrightarrow QoS_a[P_a^*] \geq \theta^* = \max_{j \in S_a} QoS_{a \to j}$$

Agent a may alter its set of predecessors by (a) adding new agents not currently in P_a, (b) changing the agents in P_a while maintaining the set size, or (c) eliminating the lowest performing predecessor(s) $i \in P_a$. Change and elimination imply selecting the lowest performing predecessor(s) i' (performance evaluation criteria depends on agent a's selection strategy) and eliminating links $fl_{i' \to a}$ and $cl_{i' \to a}$. Changes later require running NFP, while elimination requires re-configuration of the remaining predecessors through NRP. On the other hand, adding a predecessor requires only running NFP.

Figure 12.3 presents the benefit improvement protocol, a decentralized process that continuously modifies SN topology until no further improvement is possible and the SN is considered stable.

12.2   Design of Experiments

The formation/re-configuration dynamics presented in Sect. 12.1 are simulated under various agent population scenarios, and emerging SNs are challenged through various disruption modes in order to assess their resilience. Six population scenarios

Table 12.3 Population scenarios and variable levels used in simulation experiments

Scenario (no. RBT agents)	Agent population	Disruption type*	No. disrupted agents n_d
0 (baseline)	50; 100	R; T; TW	1, 2, 3, 4, 5
3	50	R; T; TW	2; 4
6	50	R; T; TW	2; 4
9	50	R; T; TW	2; 4
10	50; 100	R; T; TW	1, 2, 3, 4, 5
12	50	R; T; TW	2; 4

*R: Random; T: Targeted by degree; TW: Targeted by weighted degree

are constructed by varying the number of agents applying RBT protocols. In the baseline scenario, no RBT agents are present; SN agents are randomly assigned a predecessor selection strategy among SP—minCost, SP—maxQoS, or DP, with equal probability. Remaining five scenarios contain 3, 6, 9, 10, or 12 RBT agents, respectively, with all other SN agents being randomly assigned a strategy as in the baseline scenario. Table 12.3 presents the various levels simulated for population size, disruption type, and number of disrupted agents, for each population scenario.

In each scenario, the SN undergoes a five stage process:

(1) *Formation*: Initial population is created and predecessor selection strategies, assigned. In experiments with a population of 50 agents, 10 agents are selected as sinks (no output links) and 10 as sources (no input links). Similarly, in experiments with a population of 100 agents, sink and source agents are increased to 15. The NFP is applied by all SN agents until a network that connects all sink agents to at least one source agent is formed.

(2) *Improvement*: Once a SN is formed, agents apply the total benefit improvement protocol until stability is reached, i.e., there is no change in SN topology for Λ rounds of improvement.

(3) *Disruption*: After (2), the SN is disrupted by removing $n_{\mathbb{D}}$ agents and their links. Three types of disruptions are applied to the SN, depending on the experiment:

   i. Random: each SN agent has an equal probability of being removed
   ii. Targeted by degree: SN agents are ranked according to their number of links (degree) and the top n_d with most connections are removed
   iii. Targeted by weighted degree: SN agents are ranked according to their number of links weighted by flow (weighted degree) and the top n_d are removed

(4) *Response*: Although increased levels of situation awareness would enable agents to take preparedness actions before the impact of a disruption, it is assumed that agents operate with a basic level of SA, which allows for immediate detection of disruptions in predecessors but does not support advanced warnings. After a disruption, SN agents previously connected to an

agent that was removed are allowed to form new connections through NFP. During re-connection, agents attempt to restore their volume flow levels while maintaining, if possible, the pre-disruption QoS agreed with their successors. No cost restrictions are imposed during this stage which may lead to increased costs (versus pre-disruption) due to re-connection to restore pre-disruption flow.

(5) *Recovery*: After the SN recovers from a disruption, SN agents apply the total benefit improvement protocol until stability is reached, i.e., there is no change in SN topology for Γ rounds of improvement.

During each experiment, performance metrics are collected after stages (2), (3), (4), and (5), and topology metrics after (2) and (5). Performance is measured by total cost of flow (C_{SN}) and total quality of service (QoS_{SN}), both quantified at the sink agents A^O as follows:

$$C_{SN} = \sum_{j \in A^O} Q_{i \to j}\, C_{i \to j} / \sum_{j \in A^O} Q_{i \to j}$$

$$QoS_{SN} = \sum_{j \in A^O} Q_{i \to j}\, QoS_j / \sum_{j \in A^O} Q_{i \to j}$$

Topology is characterized by SN size (number of links), SN order (number of agents), betweenness centrality, degree, and weighted degree (by flow). Size and order provide a global description of the network in terms of the number of participating agents (out of the agent population) and the level of interconnections formed. Betweenness centrality measures the relevance of a given agent as an enabler of shortest path connectivity between any two other agents; agents with higher centrality are, therefore, more relevant to flow transmission from source to sink agents. Degree distribution characterizes the level of interconnection each agent has; for instance, long-tailed distributions indicate the presence of "hubs" that increase resilience to random disruptions but are susceptible to targeted attacks

For each experiment, 25 replications are generated using a code developed in Python 3.2.3 and NetworkX. Additional parameters for the model configuration are presented in Table 12.4.

Table 12.4 Experiment parameters

Parameter	Value/probability distribution
β_i non-RBT agents	Uniform(0.8–1.2)
$\beta_i(n)$ RBT agents	Uniform(1.2–1.6)
π_i	0.9 (high speculative behavior)
Demand of sink agents	Uniform(1500–2500) un. of flow (per agent)
Source agents parameters	QoS limit: Uniform(0.7–0.8) Cost: Uniform(7–8)—per un. of flow
No. of improvement rounds Λ	1000

12.3 Case Study Results and Discussion

12.3.1 Pre-disruption Analysis

First, the effect of RBT agents on C_{SN} and QoS_{SN} is analyzed. Simulation results for the six scenarios, with an initial population of 50 agents, are presented in Fig. 12.4. Boxplots show mean, average, maximum and minimum values of the resilience metrics, and the range between quartiles 1 and 3 thereof.

Results indicate that QoS_{SN} increases asymptotically with the number of RBT agents in the SN, despite some extreme minimum values observed for RBT equal to 9, 10, and 12. Statistical analysis with a significance level of 1% show that average QoS_{SN} increases in all RBT scenarios when compared to RBT = 0. Furthermore, a maximum increase of 16% on average QoS_{SN}, from 53% to 69%, is obtained by introducing 12 RBT agents in the SN. Results also suggest a slight increase in C_{SN} as more RBT agents are introduced; however, it is not statistically significant at 1%.

In order to test whether the effect of RBT agents on C_{SN} and QoS_{SN} changes with agent population size, experiments for RBT equal to 0 and 10 were replicated for a population of 100 agents. Figure 12.5 shows a summary comparison of the results from experiments with different population size. A statistically significant ($\alpha = 0.01$) increase of 12% on average QoS_{SN} versus the baseline scenario is found with 10 RBT agents; a result slightly lower than the 14% found for a smaller agent population size. As for C_{SN}, it is not possible to validate the apparent increase with a confidence level of 99%. Therefore, from the results obtained in experiments with different population sizes, it is possible to conclude that RBT agents can increase pre-disruption QoS_{SN} without increasing C_{SN}.

Results of significance test comparing performance of column scenarios versus row scenario (RBT = 0) are summarized in Table 12.5.

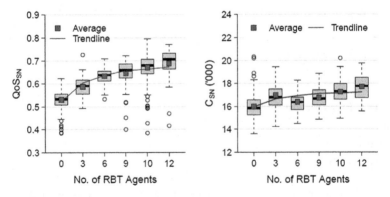

Fig. 12.4 Total QoS and total cost of flow before disruptions—population size: 50 agents

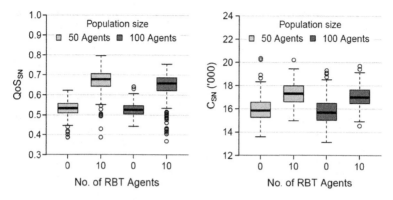

Fig. 12.5 Total cost of flow and total QoS before disruptions—population size: 50, 100 agents

Table 12.5 Supply network QoS and cost of flow—hypothesis test results at 0.99 confidence level (one sided)

Metric	Population size	No. of RBT agents				
		3	6	9	10	12
QoS_{SN} RBT = 0	50 agents	X	X	X	X	X
	100 agents				X	
C_{SN} RBT = 0	50 agents	–	–	–	–	–
	100 agents				–	

X statistically significant difference; – no statistically significant difference

12.3.2 Post-disruption Analysis: Effect on Total QoS and Total Cost of Flow

The six population scenarios were simulated under random and targeted disruptions of 2 and 4 agents, in order to analyze SN resilience. QoS_{SN} was measured after stages (2–5) to assess the level of degradation under disruptions. Figure 12.6 shows the results for simulations with a population of 50 agents. QoS_{SN} curves for all four stages measured show an upwards trend, suggesting that performance increases with the number of RBT agents regardless of the type of disruption. As expected, the impact of disruptions is higher when a larger number of agents are removed and/or when the disruption is targeted to "relevant", i.e., high degree (or weighted degree), agents. Nonetheless, the effect of targeted disruptions is reduced as the number of RBT agents increases.

Figure 12.7 shows QoS_{SN} before and after disruptions versus disruption size (no. of agents removed). Results for scenarios with a population size of 50 agents and, at least, 20% of RBT agents indicate that a SN can achieve a 15–30% increase in QoS_{SN} after a random disruption and a 10% increase after a targeted disruption (confidence level: 99%). For larger populations (100 agents) higher resilience to random disruptions than smaller networks is obtained. Nevertheless, their tolerance to targeted disruptions is reduced drastically in the case of removal by degree. This

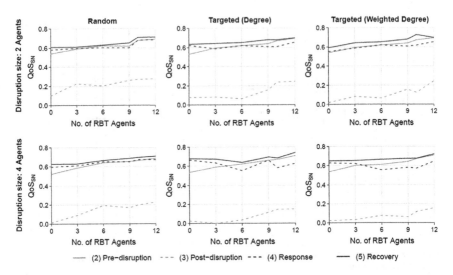

Fig. 12.6 Total QoS versus no. of RBT agents under various disruption combinations—population size: 50 agents

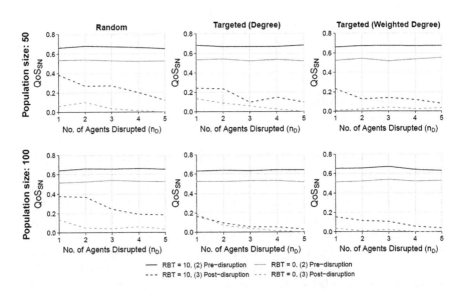

Fig. 12.7 Total QoS versus no. of agents removed under various disruption combinations

suggests that the SN formation mechanism leads to a small number of agents that concentrate a large number of links. Interestingly, the impact of agent removal by weighted degree is not reduced as much as in the previous case, also suggesting that, opposite to links, flow is not as concentrated in a small number of agents.

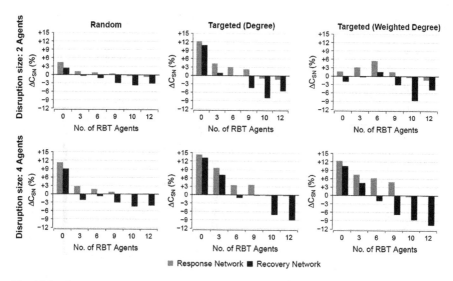

Fig. 12.8 Total cost variation versus no. of RBT agents for different disruption combinations—population size: 50 agents

Figure 12.8 compares the growth (or reduction) in C_{SN}—versus the optimized pre-disruption network—following a disruption for various levels of RBT agents. Results show that, as more RBT agents are present, lower C_{SN} increases are required to respond and recover from disruptions. Furthermore, when a sufficiently large number of RBT agents are introduced in the SN, it becomes possible to respond to disruptions without increasing C_{SN}. As for recovery, the removal of agents from the SN leads to the creation of new links and subsequent SN agent benefit improvement rounds, resulting in cost reductions (versus the pre-disruption network). These results, in combination with those presented in the preceding paragraph are a clear indication that RBT agents are capable of increasing SN resilience.

12.3.3   Network Topology: Results and Discussion

Network topology metrics were collected during the experiments, in order to characterize the structure emerging after stages (1) and (2) in the SN evolution. Two scenarios are used to evaluate the impact of RBT agents on network topology, namely RBT equal to 0 and 10.

Boxplots for SN order, SN size, and betweenness centrality (Fig. 12.9) suggest that RBT agents connect to a larger number of agents than other roles, as they are more "attractive" due to their lower cost-higher QoS combination. This, in turn, leads to SNs with approximately 50–60% more participants (order) and more than

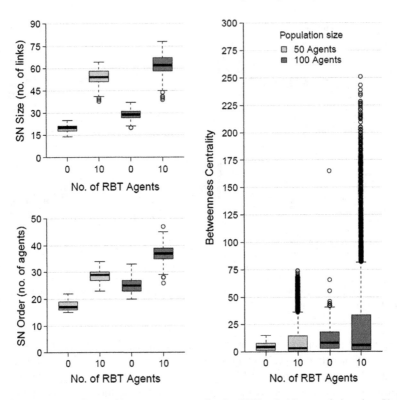

Fig. 12.9 SN order, size, and betweenness centrality for RBT = 0, 10—population size: 50, 100 agents

twice as many links (size) as in the scenario RBT = 0 (with statistical significance $\alpha = 0.01$). A higher level of interconnection partially explains the ability of emerging SNs to better sustain disruptions. Despite the removal of some agents, enough interconnections survive to deliver more flow to output agents than in the baseline scenario.

RBT agents also increase average betweenness centrality by 90–100% (with $\alpha = 0.01$) and create longer-tailed distributions. Based on the samples obtained, a Chi-square test is performed to compare distributions' shape. Using the RBT = 10 sample to obtain the probability of the expected values and the RBT = 0 sample to obtain the observed values, the test rejects the null hypothesis of equal betweenness centrality with a significance level of 1%. This behavior manifests with populations of 50 and 100 agents.

Analysis of the distribution of betweenness centrality for the scenario RBT = 10 in log scale shows that the process behind the formation of a network with RBT agents can be approximated by a power law (Fig. 12.10). Further analysis shows that the distribution is bi-modal for an initial population of 50 agents; a first mode is obtained in the lower range of betweenness centrality and a second mode, resulting

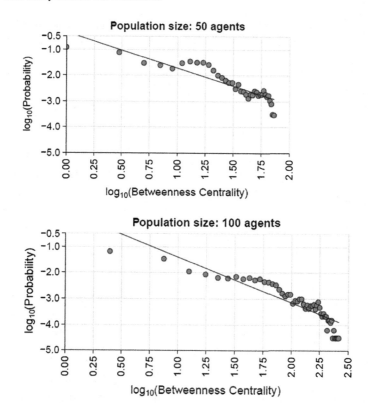

Fig. 12.10 Betweenness centrality for a SN w/10 RBT agents

from the distribution of betweenness centrality of the RBT agents, is found around 14. Following the same analysis on SNs with an initial population of 100 agents, it can be observed that the long-tail effect is heavier and predominates over the second mode introduced by RBT agents. As a result, the network topology, measured by betweenness centrality, more closely adjusts to a power law distribution.

Degree presents a behavior similar to that found for betweenness centrality. Figure 12.11 shows the degree distribution for scenarios RBT = 0 and RBT = 10, for initial populations of 50 and 100 agents. It can be observed that degree distribution for the SN with RBT agents is bi-modal in both scenarios, and that, as in the case of centrality, the second mode is a result of the addition of RBT agents. Opposite to the behavior found for betweenness centrality, degree distribution under an initial population of 100 agents still shows bi-modality.

Following the same approach used for betweenness centrality, the shape of the degree distributions for RBT = 0 and RBT = 10 were compared under the null hypothesis of equality. Based on the samples obtained from the simulations, a Chi-square test rejects the null hypothesis with a significance level of 1%, for populations of 50 and 100 agents.

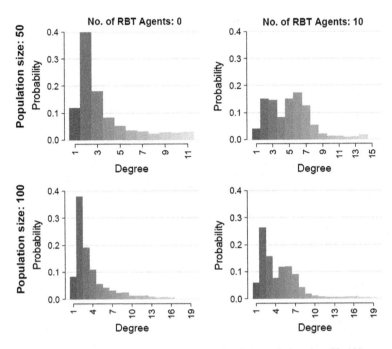

Fig. 12.11 Degree distribution for scenarios RBT = 0, 10—population size: 50, 100 agents

In the case of weighted degree, the Chi-square test also shows a statistically significant difference (confidence level: 99%) between the distributions for RBT = 0 and RBT = 10. The introduction of RBT agents produces an interesting phenomenon: a reduction on the average and standard deviation and an extension of the distribution's tail (Fig. 12.12). The reduction of average weighted degree is related to the increase in SN order generated by the introduction of RBT agents; more agents are involved in delivering the same flow to output agents. As for weighted degree standard deviation and the distribution's long-tail, the introduction of RBT agents creates a more uniform distribution of flow among non-RBT agents and concentrates volume on RBT agents. These effects combined explain the results obtained from the simulation. The phenomenon is observed with populations of 50 and 100 agents.

Results of significance test results comparing topological characteristics of SNs with RBT = 0 and RBT = 10 are summarized in Table 12.6.

The results presented in the above discussion clearly show that FTT-based design protocols can effectively influence SN formation and reconfiguration processes and alter emergent SN's topology. Section 7.2.2 shows how FTT-based design applied to individual agent-level decisions can support resilient sourcing, and the results of Case Study C show that local benefits are extensive to network level. When at least 20% of the agents in a SN use RBT protocols based on the FTT principle, emergent topology is modified so that it increases QoS in normal

Fig. 12.12 Weighted degree distribution for scenarios RBT = 0, 10—population size: 50, 100 agents

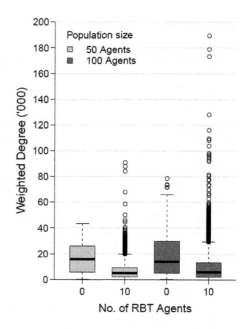

Table 12.6 Topological characteristics of SNs—hypothesis test results at 0.99 confidence level (one sided)

Metric	Population size		Statistical test description
	50 agents	100 agents	
SN order	X	X	Metric for RBT = 10 greater than metric for RBT = 0
SN size	X	X	
Average $BC(a)$, $a \in SN$	X	X	
Shape of BC (a) distribution	X	X	Metric for RBT = 10 different from metric for RBT = 0
Shape of deg (a) distribution	X	X	
Shape of weighted deg (a) distribution	X	X	

X statistically significant difference, – no statistically significant difference

operations and reduces QoS loss under disruptions, without a statistically significant increase in normal operation cost and with lower disruption response and recovery costs. These results combined allow concluding that RBT is able to increase resilience in a sustainable manner at network level based on distributed agent-level decisions. Furthermore, the generic SN used in the case study may represent a physical, digital, or service network, or any combination thereof; therefore, the results obtained are applicable to SN in general, regardless of their specific application.

References

Albert, R., & Barabási, A.-L. (2002). Statistical mechanics of complex networks. *Reviews of Modern Physics, 74*(1), 47–97. doi:10.1103/RevModPhys.74.47

Albert, R., Jeong, H., & Barabási, A.-L. (2000). Error and attack tolerance of complex networks. *Nature, 406*(6794), 378–82. doi:10.1038/35019019

Barabási, A.-L., & Albert, R. (1999). Emergence of scaling in random networks. *Science, 286* (5439), 509–512. doi:10.1126/science.286.5439.509

Brede, M., & de Vries, B. J. M. (2009). Networks that optimize a trade-off between efficiency and dynamical resilience. *Physics Letters A, 373*(43), 3910–3914. doi:10.1016/j.physleta.2009. 08.049

Ertogral, K., & Wu, D. S. (2000). Auction-theoretic coordination of production planning in the supply chain. *IIE Transactions, 32*(10), 931–940. doi:10.1080/07408170008967451

Lee, S., & Kumara, S. (2007). Decentralized supply chain coordination through auction markets: Dynamic lot-sizing in distribution networks. *International Journal of Production Research, 45*(20), 4715–4733. doi:10.1080/00207540600844050

Reyes Levalle, R. (2015). *Resilience by teaming in supply networks*. West Lafayette, IN: Purdue University.

Reyes Levalle, R., & Nof, S. Y. (2015a). Resilience by teaming in supply network formation and re-configuration. *International Journal of Production Economics, 160,* 80–93. doi:10.1016/j. ijpe.2014.09.036

Reyes Levalle, R., & Nof, S. Y. (2015b). A resilience by teaming framework for collaborative supply networks. *Computers & Industrial Engineering, 90,* 67–85. doi:10.1016/j.cie.2015. 08.017

Tangmunarunkit, H., Govindan, R., Jamin, S., Shenker, S., & Willinger, W. (2002). Network topology generators: Degree-based vs. structural. *ACM SIGCOMM. Computer Communication Review, 32*(4), 147–159.

Thadakamalla, H. P., Raghavan, U. N., Kumara, S., & Albert, A. (2004). Survivability of multiagent-based supply networks: A topological perspective. *IEEE Intelligent Systems, 19*(5), 24–31. doi:10.1109/MIS.2004.49

Chapter 13
Final Remarks and Outlook for Teaming-Based Resilience in Supply Networks

Resilience is becoming a central concern in modern complex systems. As such, supply networks need design and operation mechanisms capable of enabling multi-level resilient behavior. In order to guide researchers and practitioners in these tasks, Chaps. 2–5 introduce the basic elements required to understand resilience requirements and model SN behavior. These foundational concepts should provide the base for future research efforts in resilient systems.

Inspired in the principles of Collaborative Control Theory, Chaps. 6–9 present a teaming-driven approach to design and operate supply networks: Resilience by Teaming. A suite of protocols and design guidelines are presented in detail to provide practitioners with a generalized approach to design structures and control flow in physical, digital, and service SNs. The protocols and design considerations discussed in RBT are analyzed in real-world case studies in Chaps. 10–12.

In this final chapter, the major contributions of the book are revisited. Future research lines are discussed not only in relation to the extension of Resilience by Teaming, but also from a general perspective for complex resilient systems.

13.1 The Need for Resilience Fundamentals

Chapter 3 discusses in detail the various definitions and concepts of resilience used in engineering and systems research of physical, digital, and service networks. Despite the multiple wordings for the concept of resilience, the main semantics point to five core fundamentals that must guide the design and operation of resilient supply networks.

(i) Resilience is an inherent ability of SN agents and/or an emergent ability of supply networks

(ii) Resilience is related to the occurrence of disruptions to normal operation by undesired (but not necessarily unforeseen) events

© Springer International Publishing AG 2018
R. Reyes Levalle, *Resilience by Teaming in Supply Chains and Networks*,
Automation, Collaboration, & E-Services 5, DOI 10.1007/978-3-319-58323-5_13

(iii) Resilience involves restoring QoS to a stable, normal state
(iv) Resilience involves maintaining acceptable QoS from the occurrence of a disruption until restoration to a stable, normal state is achieved
 (v) Resilience involves active detection and prognosis of conflicts and errors potentially leading to disruptions, and implementation of preventive/preparedness actions.

The fundamentals above characterize the meaning of resilience from a conceptual standpoint. Their application in supply networks must follow a multi-level approach and have to address the two dimensions of resilience. Agents must ensure they design and operate their own resources and local networks in line with these fundamentals but also need to implement coordination and collaboration mechanisms to ensure local resilience translates into network-level resilience.

13.2   Resilience by Teaming Framework

Chapters 7–10 of this book present the Resilience by Teaming framework, developed to address the design and operation of resilient supply networks. RBT is built on three foundations: (1) a generic supply network formalism (Chap. 2), capable of describing physical, digital, and service SNs, and any combination thereof, (2) the five fundamentals of supply network resilience, presented in Chap. 3, and (3) the Fault-tolerance by Teaming principle of Collaborative Control Theory.

Existing frameworks fail to address some of the SN resilience dimensions and/or levels, or focus on balancing the trade-off between increased resilience and higher number of available resources used as protection, an approach that hinders long-term sustainability of the SN. Moreover, SN resilience frameworks found in literature often fail to operationalize the fundamentals of resilience.

On the contrary, RBT addresses SN resilience as emergent from SN agents; it encompasses a suite of design and operation protocols to enable agents to achieve local resilience while simultaneously creating the conditions for the emergence of network-level resilience. Furthermore, these mechanisms are based on a teaming approach that leverages weaker agents and can, therefore, lower the level of resources required to enable resilience. This, in turn, provides the necessary conditions for more sustainable SNs.

The main contributions of RBT are:

(1) An adaptable framework, applicable to physical, digital, and service SNs, that operationalizes the five resilience fundamentals for SNs
(2) A suite of agent-centric protocols to enable local resilience and capable to develop emerging resilient structures w/o centralized control
(3) Teaming-based structure design and flow control protocols that leverage weaker agents to produce resilient response with lower resource requirements (Table 13.1).

Table 13.1 Comparison of RBT and alternative SN resilience frameworks found in literature

Framework		Paradigm	SN Domain[a]			Dimension		Level	
			P	S	D	Structure	Control	Agent	Network
Christopher and Peck (2004)	Design principles	Trade-off	X	X		X (partially)			X
Ponomarov and Holcomb (2009)	Process (multi-step w/feedback)	Trade-off	X	X		X (partially)			X
Pettit et al. (2010)	Design principles	Trade-off	X	X		X (partially)			X
Burnard and Bhamra (2011)	Process (multi-step w/feedback)	Trade-off	X	X			X (partially)		X
Sheffi and Rice (2005)	Process (8 phases)	Trade-off	X	X		X (partially)	X (partially)	X	X
Sterbenz et al. (2011a, b)	Process (6 phases in 2 cycles) based on 4 resilience axioms	Trade-off			X	X	X (partially)		X
Smith et al. (2011)	Process (4 phases w/feedback)	Trade-off			X		X		X
Jackson and Ferris (2013)	Design principles	Trade-off	X	X	X	X (partially)			X
Resilience by Teaming	Resilience fundamentals + FTT applied to design and control protocols	Teaming	X	X	X	X	X	X	X

[a]P physical, S service, D digital

The abovementioned contributions are significant towards enabling sustainable resilience in the following aspects:

(1) Under certain conditions, parallel processing by a team of weaker resources in IRNs can outperform a single, highly reliable resource (as proven in Sect. 8.1.1). This result enables IRN design not based on resilience vs. cost trade-off.

(2) Use of teaming-based IFCP in IRNs subject to random failures increases QoS by 0.45% and reduces throughput variability by 43% while simultaneously lowering the level of resources (WIP) required by 24% (vs. traditional control approaches).

(3) Teaming-based (DNF/RP + DFCP) distribution of flow subject to congestion and disruptions performs as well as protocols optimized for shortest-time delivery but with a 4.5% cost reduction. Furthermore, teaming-based control protocols can adapt to various levels of congestion, maintaining the performance advantages vs. traditional methods.

(4) RBT enables agent-level resilience by teaming weaker agents/resources (as shown by (1), (2), and (3) above) but also fosters emergent supply network resilience. Case Study C shows that, when at least 20% of agents operate under RBT, Q_oS_{SN} increases during normal operations and is better preserved during disruptions (10–30% lower losses post-disruption, depending on the disruption type). Furthermore, these network-level enhancements are achieved without increasing cost, thus enabling SNs to be more sustainable in the long term.

Resilience by Teaming, RBT, advances current knowledge of resilience-enabling protocols based on teaming, an approach more capable of ensuring long-term SN sustainability than alternatives based on trade-offs between increased use of resources and resilience.

13.3   Teaming-Based Resilience in Supply Network Design: Open Questions

As outlined by the dimensions of supply network resilience in Sect. 3.3, there is an interplay between SN structure and control protocols. In this sense, the former significantly define the ability of the latter to enable resilient performance, both at agent and network levels. In order to provide the right conditions for teaming-based flow control protocols to achieve high resilience with low resource consumption, Sects. 7.2, 8.1, and 9.2 define the mathematical foundation and protocols to design resilient structures using weaker agents.

In RBT, structural design has been addressed from an agent focus, establishing protocols and guidelines for teaming-based design. STF/RP defines how to size and select participants for a primary sourcing team P_a^1, taking into account possible topological and topographical effects on disruption correlation. The approach guarantees that required delivery leadtime Δt_r during flow control can be satisfied with pre-defined probability θ^*. P_a^1 is coupled with a secondary team P_a^2, responsible for sourcing an agent the $1 - \theta^*$ fraction of the time where P_a^1 cannot meet Δt_r. Combined, and controlled by SFCP, P_a^1 and P_a^1 are able to provide stable sourcing to agent a, despite any disruptions affecting participants thereof.

As formalized in Sect. 2.2.1, each SN agent comprises a set of internal resources R_a which are arranged in an internal network, IRN, and are responsible for transforming input flow into output flow. The input/output transformation is required to meet pre-defined levels of QoS, and must be able to match output demand with minimum variability. To this end, the design of processing stages and storages is addressed from a teaming perspective. Section 8.1.1 provides mathematical proof that, under certain conditions, a processing stage composed of parallelized weaker resources can outperform a single resource with high reliability, from a throughput variability perspective. Section 8.1.2 establishes the design criteria to size storage spaces when operated through a teaming-based protocol, e.g., IFCP. Adaptive use of storage capacity based on prognosed evolution of the IRN enables a reduction of storage space requirements to achieve resilient performance.

SN agents' responsibilities include sourcing, internal processing, and flow distribution, all of which require design for resilience. In order to address resilient distribution of flow, Sect. 9.2 introduces DNF/RP, a protocol for smart, dynamic selection of delivery paths to anticipate and overcome congestion and disruptions. Based on real-time data regarding flow links and intermediaries, DNF/RP adaptively selects an optimal path given available time to deliver a flow order and structural flexibility provided by the intermediaries' network. DNF/RP's effectiveness is tested in Case Study B, where combined with DFCP achieves a performance equal to shortest-time delivery protocols but with 4.5% lower costs of operation.

Teaming-based design driven by local decisions can also affect the topology of emerging supply networks. As shown in Case Study C, the addition of RBT agents effectively modifies SNs' topology, creating a more interconnected nucleus of kernel agents. Results indicate that RBT agents (1) influence the formation process to create SNs with higher order and larger size, (2) provide higher interconnection levels which affect betweenness centrality by increasing its average value and creating a long-tail effect, (3) introduce a degree distribution with higher average value which, in turn, produces a bi-modal degree distribution (when all agents are considered), and (4) reduce the average weighted degree and create a long-tail effect. These effects on network formation lead to increased resilience under normal and disrupted modes, using less costly resources and, therefore, enabling sustainable long-term operations.

Various questions related to design of SNs based on teaming remain open. As proven in Case Study C, RBT can effectively increase resilience in SNs; however, there are no mechanisms to ensure that resilience gains are fairly distributed among all SN agents (as opposed to being concentrated on a few). Distributed allocation of protection through storage and excess capacity/parallelism also requires further analysis to ensure it is optimally allocated from a network perspective but also capable of providing fair protection to all agents in the SN. To this end, protocols that enable sharing of protection capabilities need to be developed, possibly inspired by demand and capacity sharing (DCS) protocols.

13.4   Teaming-Based Protocols for Resilient Supply Network Operation: Open Questions

Inspired by the Fault-tolerant Time-out Protocol (Liu and Nof 2004), an application of FTT to sensor networks, Sect. 7.3 introduces SFCP, a control protocol which generalizes insights gained from FTTP. Following the design of sourcing teams P_a^1 and P_a^2, SFCP effectively teams agents in P_a^1 to leverage their weak performance. The protocol is designed to accept sourcing deliveries that would be disregarded by traditional approaches focused only on delivery leadtime requirements. Furthermore, SFCP creates dynamic protection by allowing some level of overlap among deliveries, which act as transient safety stocks. P_a^2 provides an additional layer of response to potential disruptions; SFCP engages in flow sharing with agents in P_a^2 whenever flow cannot be sourced from agents in P_a^1.

Flow received from predecessors requires internal processing to be transformed into output flow with the characteristics requested by an agent's successors. Traditional protocols for IRN control lack the required degree of anticipation, recovery, and response capabilities to cope with resource failures in an IRN. IFCP, introduced in Sect. 8.2, is designed to enable teaming among resources in an IRN to overcome failures and prevent QoS loss and variability. Inspired by the FTT and CEDP principles of CCT, IFCP integrates early conflict detection capabilities with adaptive protection to maximize the use of resources towards avoiding or reducing the impact of possible failures on IRN's QoS and throughout variability.

Results from Case Study A validate the above-mentioned observations regarding IFCP. When compared to the next best performing control protocol, IFCP achieves a 0.45% increase in QoS while simultaneously reducing WIP by 24% and throughput variability by 43%. The reason for this difference lies in the adaptive method's ability to adjust to different system states by updating target levels or job release rules according to information obtained in real time. Furthermore, it can be concluded that models which lack anticipatory capabilities, cause higher throughput variations when compared to IFCP. This difference can be explained by higher disturbances occurring after a failure when actions are enforced to stabilize the system and maintain throughput. Results also show that anticipation capabilities and conflict prevention protocols are required to obtain reductions in WIP by minimizing failure interactions without compromising throughput.

Finally, control of resilient distribution of flow is obtained by DFCP. Section 9.3 introduces a protocol that dynamically re-configures delivery paths based on updated network conditions. Through DFCP, intermediaries team-up to ensure timely delivery of flow, with minimum cost and need for contingent/preventive re-routing. Results from Case Study B show that fuzzy rules are able to adequately capture the trade-off between cost and time while also incorporating the notion of path flexibility as a means to favor low-cost, lower-performing paths that can be later abandoned if congestion and/or disruptions arise. This dynamic path (team) formation and re-configuration enables low-cost resilient flow delivery which, in turn, contributes to sustaining operations in the long term.

Teaming-based flow control in SNs presents several opportunities for further exploration. In the case of SFCP, parameterization requirements for various SNs need to be explored and the resulting agent-level benefits need to be further quantified. As for IFCP, its ability to leverage available resources through teaming depends, to a great extent, of the capacity of ECDT to accurately detect possible disruptions occurring in the future. Therefore, application-specific conflict detection tools need to be further explored. Additionally, adaptive learning capabilities need to be embedded into IFCP, in order to provide it with automated responsive re-parameterization upon context changes in the IRN. Finally, DFCP requires further analysis of rules and conditions to better cope with high-congestion scenarios.

13.5   Future Research Directions

Resilience research is quickly gaining momentum. As SNs become increasingly complex and with growing level of interactions, often hidden or hard to anticipate, mechanisms that can ensure efficient operation under normal conditions and effective and sustainable operation when challenged by disruptions are required. To this end, future work directions and extensions of RBT to address its current limitations include:

(1) Exploration of DCS-inspired protocols to enable fair allocation of resilience-enabling resources and performance benefits among agents.
(2) Re-visit of various existing methodologies, e.g., lateral transshipments, from a teaming-based perspective and analyze their performance at network level.
(3) Testing of sourcing algorithms in various settings: digital, physical and service networks. Validation of previous conclusions of application-specific protocols such as FTTP.
(4) Design of autonomous learning systems to dynamically adjust protocol parameters and teaming rules, and anticipate to possible changes in SN structure/function. This minimizes human intervention and enables smart scenario-analysis to fine-tune control protocols and increase, even in the face of system changes.
(5) Develop protocol extensions capable of managing SLAs with different resilience priorities.

(6) Analysis of the effect of communication network topology on situation awareness and, consequently, SN resilience.
(7) Evaluation of the impact of different situation awareness mechanisms and levels (at agent- and network-level) on the ability of agents and the SN to anticipate and prepare for disruptions, and, therefore, increase resilience.

References

Burnard, K., & Bhamra, R. (2011). Organisational resilience: Development of a conceptual framework for organisational responses. *International Journal of Production Research, 49*(18), 5581–5599. doi:10.1080/00207543.2011.563827

Christopher, M., & Peck, H. (2004). Building the resilient supply chain. *The international journal of logistics management,* 15(2), 1–14. doi:10.1108/09574090410700275.

Jackson, S., & Ferris, T. L. J. (2013). Resilience principles for engineered systems. *Systems Engineering, 16*(2), 152–164. doi:10.1002/sys.21228

Liu, Y., & Nof, S. Y. (2004). Distributed microflow sensor arrays and networks: Design of architectures and communication protocols. *International Journal of Production Research, 42* (15), 3101–3115. doi:10.1080/00207540410001699363

Pettit, T. J., Fiksel, J., & Croxton, K. L. (2010). Ensuring supply chain resilience: Development of a conceptual framework. *Journal of Business Logistics, 31*(1), 1–21. doi:10.1002/j.2158-1592. 2010.tb00125.x

Ponomarov, S. Y., & Holcomb, M. C. (2009). Understanding the concept of supply chain resilience. *International Journal of Logistics Management, 20*(1), 124–143. doi:10.1108/ 09574090910954873

Sheffi, Y., & Rice, J. B., Jr. (2005). A supply chain view of the resilient enterprise. *MIT Sloan Management Review, 47*(1), 41–48.

Smith, P., Fessi, A., Lac, C., Hutchison, D., Sterbenz, J. P. G., Scholler, M., et al. (2011). Network resilience: A systematic approach. *IEEE Communications Magazine, 49*(7), 88–97. doi:10. 1109/MCOM.2011.5936160

Sterbenz, J. P. G., Cetinkaya, E. K., Hameed, M. A., Jabbar, A., Qian, S., & Rohrer, J. P. (2011a). Evaluation of network resilience, survivability, and disruption tolerance: Analysis, topology generation, simulation, and experimentation. *Telecommunication Systems, 52*(2), 705–736. doi:10.1007/s11235-011-9573-6

Sterbenz, J. P. G., Cetinkaya, E. K., Hameed, M. A., Jabbar, A., & Rohrer, J. P. (2011b). Modelling and analysis of network resilience. In *Proceedings of the third international conference on communication systems and networks* (pp. 1–10). Bangalore, India. doi:10.1109/ COMSNETS.2011.5716502

Printed in the United States
By Bookmasters